CW01572406

POLITICS AND NEO-DARWINISM

And Other Essays

TOM RUBENS

SOCIETAS

essays in political
& cultural criticism

imprint-academic.com

Published in the UK by Societas
Imprint Academic, PO Box 200, Exeter EX5 5YX, UK

Published in the USA by Societas
Imprint Academic, Philosophy Documentation Center
PO Box 7147, Charlottesville, VA 22906-7147, USA

ISBN 9781845402495

A CIP catalogue record for this book is available from the
British Library and US Library of Congress

Contents

III

Prefatory Note

The following series of essays falls into three main sections. The first section, essays 1–8, examines a range of political, ethical, and cultural issues. The second, essays 9–15, is chiefly cultural, and focuses on certain general aspects of the Western intellectual context. The third, essays 16–30, concentrates on strictly philosophical issues. While similar to material in section 2, it looks in more detail at specific philosophical questions.

T.R., August, 2010.

Politics &
Neo-Darwinism

It should straight-away be specified that this essay is not about those systems of thought which go under the heading of 'social Darwinism'. Its subject is the relationship between politics, considered essentially as a moral activity, and neo-Darwinism, the doctrine which is the norm in modern biology and which, for most modern biologists, constitutes a secular-naturalistic view of man. The focus, then, is on the connection between moral values, when politically channelled, and secular naturalism.

A useful observation to begin with is that, in the West over the past two centuries, politics has become increasingly secularized.[1] There are a number of reasons for this, but prominent among them have been the advance of empirical science and, concomitantly, the decline of religious belief.

This is of course not to say that there has been no linkage at all between political activity and religious commitment. The Christian Democrat parties in Germany and Italy are instances of overt and pronounced connection between the two contexts, as are the Neo-Conservatives in the United States. But these are exceptions, and in general the relation has lost most of the emphasis and definiteness it once had, or at least professed to have, especially in mediaeval times. For many generations now, it has been rare for political projects to refer extensively to religious culture in order to validate themselves—again, unlike earlier periods.

---

[1]   This process has not of course been confined to Western politics, as is indicated by, for example, the political history of the Soviet Union from 1917–89, and of China since 1949. But in these cases, the secularisation resulted from the influence of a *Western* philosophical doctrine: Marxism.

Given, then, the increasing disconnection between politics and religion, but given also the moral role[2] which religion, at its best, has traditionally played in human culture, the question arises: what essentially ethical function can a secularised politics have?

The question carries two implications. One is that politics should be, to repeat, a moral activity. The other is that this activity should be undertaken within a secular-naturalistic perspective; and part of this perspective is rejection of the view that moral values can be discovered by or revealed to mankind, as entities existing outside the human mind.

Taking the first of these two implications: politics as ethical commitment means, as does every other kind of such commitment, an effort to maximise the well-being of the group, through harmonising individual interests and extending mutual understanding and reciprocity. Ideally, the group should be the entire human race. Thus political projects should seek, as far as possible, to be global rather than merely national or regional.

This aspiration will inevitably encounter formidable obstacles. Chief among these are vested economic interests, and the political apparatus which goes with them: economically dominant groups within nations and across nations, and the influence they exert, in various ways, on government in both its national and international forms, i.e. on the national state, and on inter-state organisations (such as the European Union). Other obstacles include racial, ethnic, nationalistic, and religious prejudices: factors generally not as powerful as economic ones but still carrying enormous weight. These difficulties must be seen as ongoing.

Clearly, a truly altruistic global politics can only be conducted by people who stand above vested interests and sectarian viewpoints of any kind: those who are certainly concerned about the welfare of their own nation (the whole of it) but are equally concerned with that of all nations. In this sense, they love their neighbours as themselves; and

---

[2]   Indeed, its only meaningful role, most secularists would argue.

they should remain undeterred even if they find that this love is not the norm in the world.

Moving now to the second implication: the secular-naturalistic perspective rules out ethical objectivism, the view that moral values are discoverable by the human mind. It argues that mankind *makes* these values and does not find them. Humanity creates them but does not encounter them. Thus the position is ethical inter-subjectivism.

This position means that, often, tough political/moral decisions will have to be made, some involving extensive suffering and death; and made in the conviction that they cannot be validated by reference to anything outside the human mind — to, for example, ideas of deity and of deity-revealed concepts of right and wrong. The unavailability of such vindication can produce a sense of moral anguish: a situation which is, incidentally, nowhere better described among modern writers than by Sartre. All participants in a politics based on secular-naturalism must continually remember that no extra-human validation of their actions is possible for them. To forget or evade this would be to fall into what Sartre aptly calls "bad faith".

Reference to modern authors such as Sartre can be extended to include Thomas Mann. Mann famously said, "In our time, the destiny of man presents its meaning in political terms". Leaving aside the complications attached to the use of the word 'destiny', we can recognise the obvious magnitude of this statement, involving as it does a very general deployment of the term 'political', to mean, basically, everything connected with mankind's social values, arrangements, and actions. Thus the burden of the statement is that the future of humanity will be the product of socially collective action undertaken now. Given the manifest relevance of this position to current issues such as climate change, over-population, and the diminishing of natural resources, it is clear that nothing could be more in agreement with the politics of secular-naturalism, and with the neo-Darwinian perspective as a whole.

Shakespeare & Sartre: The Defence of Political Violence

I. Shakespeare

There can be no doubt that Shakespeare morally endorsed the use of political violence under certain circumstances. The ethical role that such violence occupies in his work is pivotal and recurrent, as much in the tragedies as in the history and Roman plays. Stressing this function is important for a number of reasons.

Firstly, the fact that the role is not confined to the history plays must be considered in the light of the argument, as advanced by some critics, that these plays are merely an exercise in endorsing the Tudor view of history, and are therefore just propaganda. Even if it is true that Shakespeare did write them with a pro-Tudor attitude, and did deliberately distort a number of historical facts, the resultant material is not propaganda. The latter specialises in moral and psychological oversimplifications, and these are totally absent from the history plays, which unquestionably display the complexity of real life. Hence, though the plays may not be true to certain particular historical facts, they are definitely true to life as a whole. Further, they do mirror the *general* historical and cultural context in which they were set; they convey an authentic sense of an overall quality and way of life. They are, then, in some ways a highly realistic fictional alternative to what actually happened. As such, they possess a moral framework which deeply addresses perennial ethical concerns.

In this framework, certain forms of political violence are presented as justified and praiseworthy. Examples include Richmond's military defeat of Gloucester at the end of *Richard III,* and Henry's invasion of France and victory at Agincourt in *Henry V.* Whatever fictional distortion there may be in these and other plays, we feel that *if* circumstances in real life had been what they are presented as, then the violence in question would have been justified.

Also, given that the history plays are set in the Middle Ages, it should be noted that Shakespeare, in so far as he did adhere to Christian values, did not subscribe to those parts of original Christian doctrine, as found in the Gospels, which reject the use of violence. In his acceptance of violence, he reflects the militant aspects of mediaeval Christianity, ones very different from those in the Gospels, and so the cultural heritage which this later form of Christianity bequeathed to the Elizabethan period.

In close association with the act of violence is the threat to use it; the power of deterrence was equally a part of Shakespeare's political thinking. For instance, in *Henry VIII,* he speaks of a thriving politico-social system as one which contains "Peace, plenty, love, truth, *terror* [italics mine]" (I, 3, l.26). The inclusion of terror with other, different virtues is evident also in one of the non-history plays, *The Winter's Tale,* where terror is coupled with, again, love (I, 1, l.19).

As regards the Roman plays, attention will be confined to *Julius Caesar*, arguably the most philosophical of this group of dramas, and the one that goes into considerable detail on the ethics of political assassination. Brutus's closely-reasoned arguments, both before and after the assassination, for the legitimacy of killing Caesar in order to prevent dictatorship, are set against Mark Anthony's eulogy to Caesar, but are not actually refuted by the latter. Anthony stresses Caesar's generosity to the people, and his refusal of a kingly crown. Anthony claims that these gestures showed lack of ambition; but it is arguable that they showed, not this, but strategic sagacity on the part of someone who actually was ambitious.

Further, Anthony is clearly intent on psychologically manipulating his audience, whereas Brutus is not. So, Brutus's ethical reasoning remains a cogent force in the play, and his weaknesses are presented as lying in the sphere, not of morality, but of political and military practice (a sphere in which Anthony, by contrast, is strong). Hence we can say that the ethics of political assassination is a central issue in the play, and is so because Shakespeare did *not* rule out such action as a matter of principle.

Turning to the tragedies: the most obvious case of justifying political violence is surely to be found in *Macbeth*. Macbeth's murders of King Duncan, of the two chamberlains, of Banquo, and of MacDuff's family, plus the fact that he exerts tyrannical power as monarch, create a situation which, in the eyes of the audience as much as in those of the characters in the play, calls for military action against him. This action, along with Macbeth's refusal to surrender, inevitably leads to the tyrant's slaying. The latter event produces no regrets (even though the audience remains mindful of the overall tragedy of Macbeth's moral decline).

Less prominent but still crucial is the role of violence in resolving the situation in *King Lear*. The French army, led by Cordelia, tries but fails to defeat the forces led by those who are guilty of treachery and ruthlessness. However, the latter are finally defeated, through the outcome of single combat between Edgar, the force for good, and Edmund, traitor and self-seeker. Without Edgar's fighting skills, this defeat would not have been achieved.

In *Othello,* the Moor as tragic protagonist begins the play as, like Macbeth, a highly respected military man, honoured for his prowess in defence of the Venetian state. Shakespeare gives no suggestion whatever that we, the audience, should have a less positive attitude toward Othello, because of his use of violence, than do the Venetian statesmen. This implicit endorsement of physical force is reinforced by the considerations of naval strategy against the Turks which occupy a sizeable space in the opening part of the play.

In *Hamlet,* matters are more complicated, in the sense that the central character lacks a clear-cut response to the injunction placed on him to perform an act of physical violence. Without going into the immense psychological and thematic complexities of the play, it will be sufficient, for present purposes, to make the following observations: 1) Hamlet does, as enjoined, eventually kill the usurping king. Because the latter has remained relentlessly ruthless since the murder of the elder Hamlet, his death is generally regarded by audiences as an event to be viewed positively. In this sense, Hamlet's action can be seen as a definite achievement. 2) Hamlet, at least with one part of his mind, has wanted to kill the king ever since the revelations to him in Act I. This persisting psychological fact, in a man who is morally upright, can be read as a sign that, again, Shakespeare regarded violence as an ethically viable option.

3) Hamlet has previously displayed overt willingness to use violence, when the ship on which he was travelling to England was attacked by pirates. Also, he evinces an approval of violence being used by others when he arranges for Rosencranz and Guildenstern to be executed in England, instead of himself. Hamlet's endorsements of physical force more than suggest a link between his attitude and Shakespeare's.

One of the many reasons Shakespeare remains an author of seminal relevance is his acceptance of the need for political violence in certain situations. For the majority of people in the Allied nations who lived through the most cataclysmic series of events of modern times—the Second World War—this need was self-evident. It is so whenever the concept of a just war is operative. To engage in violence is of course to risk death in the name of a cause, and whatever the ontological framework in which death is viewed, be it secular or religious, willingness to pay the mortal price is an inevitable feature of such engagement.

This range of considerations is present, implicitly or explicitly, throughout much of Shakespeare's work, but perhaps nowhere more vividly than in Henry V's St. Crispin's Day oration, before the battle of Agincourt, on military comradeship in the face of death, and on the possibility of gaining enduring honour in pursuit of a just cause:

> If we are mark'd to die, we are enow / To do our country loss; and if to live, / The fewer men, the greater the share of honour… He that outlives this day, and comes home safe, / Will stand a tip-toe when this day is nam'd / And rouse him at the name of Crispian… And Crispin Crispian shall ne'er go by, / From this day to the ending of the world, / But we in it shall be remembered; / We few, we happy few, we band of brothers…[1] (*Henry V,* IV, 3, ll.20–60 [extracts])

II. Sartre

Reference to WWII inevitably leads us to a consideration of Sartre, who was a member of the Resistance during the Nazi occupation of France from 1940–44, and for whom the war was a morally formative period because it provided him with his first opportunity for fully-fledged political commitment. Whereas with Shakespeare debate is possible about his degree of adherence to Christianity, with Sartre no debate is needed. He was an unequivocal atheist, and this determined the entire framework of his thinking about political violence.

That framework was a complex one, and is to some extent captured when he speaks of "affirmations of systems of values and rights such as the rights of citizenship, the rights of the family, individual ethics, collective ethics, *the right to kill* [italics mine]".[a] Note how the last in the list of rights is seen as part of a larger network of social and political considerations; and we can be sure that the pursuit

---

[1]   Given the previous reference to WW2, it is worth noting that this speech was one of the dramatic high points of the 1944 film version of *Henry V,* directed by, and starring, Laurence Olivier. The credits of this film begin with a dedication to the Allied forces who took part in the D-Day landings, June 6th, 1944.

of all these rights actuated Sartre in his work for the Resistance, as well as in his subsequent political activities, which were extensive.

However, his endorsement of political violence is most widely known, not through his political efforts but through his literary work: chiefly, his plays *The Flies, Crime of Passion, Lucifer and the Lord,* and his novel *Iron in the Soul.*

The Flies (1942) was Sartre's first play, dealing on one level with the Orestes story as derived from ancient Greek drama, but it is also an allegory of the contemporaneous situation in occupied France. In rendering his own version of the narrative of Orestes's killing of Aegisthus — the man who was the murderer of Orestes's father Agamemnon, also the lover of his mother Clytemnestra, and illicit co-ruler of Argos — Sartre explores many issues. Several of these are to do with Orestes's own personal situation, but some are explicitly collective and political in character. Orestes's slaying of Aegisthus has public as well as private dimensions, and the play avers that "An evil thing is conquered only by another evil thing", i.e. the violence of Aegisthus must be met with counter-violence. The latter is eventually delivered by Orestes, who regards his action as a moral and political coming-of-age. The broad implication is that the French people must take violent action against their Nazi oppressors.

In his 1948 play, *Crime of Passion,* Sartre strongly echoes Shakespeare's *Julius Caesar* by exploring the issue of political assassination. The most sympathetic character in the play, the Communist leader Hoederer, holds the view that in political struggle the end justifies the means. Therefore violence is justifiable as a means: "My hands are filthy. I've dipped them up to the elbows in blood and shit." Thus Hoederer endorses political assassination where deemed necessary.

In considering Hoederer's points, we should not lay undue stress on the fact that he is a Communist. His acceptance of violence is an attitude which transcends specific political parties and historical locations; it pervades the whole of political history—a point which is surely clear from the study of Shakespeare. However, if we did want to be historically specific, we could look again at WW2, and point out that the Allies laid plans to assassinate several leading Nazis; only circumstances, not moral compunctions of any kind, prevented the realisation of these plans.

To a significant extent, Hoederer's views are echoed in Sartre's 1952 play, *Lucifer and the Lord.* Set in Germany at the time of the Lutheran Reformation, this focuses on the character of Goetz, an ex-nobleman who decides to try and free the peasantry from the domination of the barons. Initially, he thinks he can achieve this goal solely by practising the Gospel virtues. But experience painfully teaches him that only returning good for evil, seeking peace, showing love and being non-violent, are ineffective. He realises he must be violent and ruthless toward the enemy because "To love a man is to hate the same enemy". Also— and in this we hear Hoederer's voice:

> Men of today are born criminals; I must demand my share
> in their crimes if I desire my share of their love and virtue…
> Good and evil are inseparable. I accept my share of Evil to
> inherit my share of Good.

Once again, support for this position comes from history in general, from all instances of prosecution of a just war.

Finally, let us consider Sartre's 1949 novel, *Iron in the Soul.* The third in a trilogy called *Roads to Freedom,* this novel climaxes in an act of political violence performed by Mathieu, the trilogy's central figure. Previously, Mathieu has mainly been an indecisive, hesitant figure, given mostly to self-analysis rather than action (not unlike, in fact, Shakespeare's Hamlet). But, with the invasion of France by the Nazis, his powers of decision are activated, and he joins a rural militia in an attempt to delay the Nazis' advance. All the odds are against success in this venture, but still Mathieu continues to fire his rifle, in his now unflagging commitment to action. Like Goetz in *Lucifer and the Lord,* he is keenly aware that his determination to kill violates Gospel morality: "He fired, and the tables of the Law crashed about him — Thou Shalt Not Kill — bang! at that scarecrow opposite." Through this violence, Mathieu regards himself as "cleansed" and "free". Once more, we the readers are invited to empathise with the person who kills from a sense of justice which breaks with moral orthodoxy.

To conclude: over 300 years separate Shakespeare from Sartre, yet, despite major differences in historical context, their defences of political violence are essentially the same. This fact leads us to think that such defence will again be the same in 300 years time, when a future writer of intellectual stature offers it. Whether such a writer is non-religious or religious, the practical issues will be unchanged: the need to take up arms, await opportunities and, when the latter come, act decisively. Preparedness to shed blood remains what it is no matter which general ontological outlook that preparedness is part of.

[a]   As quoted by Allardyce Nicoll in *World Drama*: London, Harrap and Co. Ltd., 1959 (1949), p. 908.

Economic Hardship in Shakespeare's Plays

The previous essay examined a key aspect of Shakespeare's political outlook. This outlook, it should be added, incorporated a very wide range of political concerns. Shakespeare was very much a political writer, intensely interested, from an ethical standpoint, in affairs of state. This meant a focus on actions and events at the top level of society, where political power lay. Such a focus entailed, however, a relatively brief treatment of material realities at society's lower levels, where economic hardship and indeed poverty were widespread.

This latter observation is not necessarily a criticism of Shakespeare. Other great writers too have paid comparatively scant attention to issues of hardship and poverty: for example, Homer, Aeschylus, Virgil, Racine, Austen, Proust. Yet they are no less great for this omission. Clearly, literary eminence depends on what a writer does with the particular material s/he chooses to deal with. At the same time, the observation about Shakespeare does imply some re-valuation of him: certain other great writers *have* explored economic privation to a far larger extent—for instance, Fielding, Dickens, Hardy, and Zola. It is to writers such as these, rather than to Shakespeare, that economic historians can turn for a more inclusive picture of the past's material realities.

Of the last four authors named, it will be seen that three belong to the 19th century. Also, it should be noted that most of the authors who have extensively concerned themselves with the overall economic situation belong to the 19th and 20th centuries. In this regard, it is significant that the last 200 years have seen, in the West, the large-scale growth of political democracy and of the democratic outlook in general. During this period, Western society has largely lost its traditional aristocratic framework, based as the latter was on a feudal, pre-commercial, and pre-industrial kind of social organisation. The expansion of democracy has, broadly speaking, gone hand in hand with the development of a commercial and industrial form of society.

The aristocratic framework which these developments have replaced can be viewed as at least a partial explanation for the limited extent of interest in the economic totality displayed by so many — perhaps most — writers prior to the 19th century. In these earlier times the literary imagination was mainly oriented toward those who held powerful and privileged positions in society, and/or those who displayed exceptional and heroic qualities as individuals. In this way, the perspective was a distinctly selective one. Thus general economic considerations were marginalised. This perspective appears to have included Shakespeare. There is very good reason to think that his social outlook was hierarchical: hence his predominant political interest in what transpired at the top level of society.

Nevertheless, there are points in his writing which do evidence concern with the economic totality, and especially with hardship as a general social fact. Significantly, these points are found in the latter half of his career, from *Hamlet* onwards, and particularly in *King Lear.* It was in this period, and chiefly in the four great tragedies which include *Hamlet* and *Lear,* that Shakespeare evinced his deepest insights into the general human condition — which includes, of course, an economic condition.

It is true that in a pre-*Hamlet* play, *Henry V,* he does make detailed reference to the harsh working conditions of rural labourers; but this is mainly to draw a contrast between the unbroken sleep they enjoy at night, as a result of their physical exertions, and the broken and troubled sleep of kings, who bear all the psychological burdens of power.

Hence it is not clear that this reference is an indictment of the economic situation it describes.

However, by the time of *Hamlet,* Shakespeare appears to be unequivocally critical of that situation. The Prince speaks, with his characteristically panoramic outlook and compassion, of those who "grunt and sweat under a weary life" (III, 1). Further, in *King Lear,* we have a situation where two old men from an aristocratic background, Lear and the Duke of Gloucester, directly experience economic destitution (unlike Hamlet); and, as a consequence, come to an understanding of the general plight of the poor: an understanding which they have never previously even approached. Both advocate that the wealthy should give what they do not strictly need to the poor. As Gloucester says, "So should distribution undo excess / And each man have enough" (IV, 1). Given what has been said about Shakespeare's probably hierarchical outlook, such advocacy can be interpreted as showing his endorsement of a social stratification which, while firmly delineated, is humane and just. Society should have a top level, but the latter should not be excessively affluent.

Overall, *King Lear* presents a more detailed picture of material privation than is found in any other Shakespeare play. The privation has a rural as distinct from urban context; and, since English society in and prior to Shakespeare's day was mainly agrarian, the detail has a representative status. We see another aristocratic figure, Gloucester's son Edgar, forced into a fugitive and destitute state, and deciding to disguise himself as one of the many beggars who wander the countryside seeking charity from lowly farms, small villages, sheep-cotes, and mills: that is, from people who can scarcely afford to give charity. Another glimpse of rural poverty comes when an old man leads Gloucester after the latter has been blinded. This man has been Gloucester's land-tenant for eighty years; he tries the best he can, but can offer very little material support. Finally, a brief indication is given of how desperate some men are to escape poverty. When Gloucester's other son, Edmund, becomes army leader and offers his captain the chance to advance himself by committing murder, or else completely lose the status he has gained so far, the captain accepts. "I cannot draw a cart nor eat dried oats", he declares (V, 3), referring to the dismal economic condition he would otherwise have to (re)turn to.

All these details about the economic situation contribute in no small measure to the general picture of environmental bleakness which the play conveys. Few other dramas present such a strong sense of mankind as surrounded by huge, alien and comfort-less physical space: the space in which human beings have to survive, and to which the poor, with their inadequate shelter and clothing, are the most exposed.

For us, looking back 400 years to the economic conditions which the play depicts, there is of course a sense of the enormous material and social progress which has since been made in English society, and in Western society as a whole. At the same time, we are conscious that the rural conditions shown in *Lear* have their counterparts today in large areas of the Third World, where society is still predominantly rural, and where many rural economies are still feudal or semi-feudal. The sympathy which Shakespeare, briefly but unforgettably, showed for those subject to hardship in the England of his day should find an equivalent in us: in the attitude we should have toward people anywhere in the world where large-scale privation is continually suffered.

Intellectual Foci in British Society since 1945

It is instructive to consider the different kinds of intellectual focus evident in British society over the past 64 years. 1945 has been chosen for obvious reasons. It marks both the end of World War Two, which had a greater impact on this country, as on many other countries, than any previous war; and the beginning of a social phenomenon unprecedented in British — indeed world — society: the Welfare State.

As regards the impact of World War Two: it weakened Britain's military position in the world to such an extent that almost all of the British Empire, the most extensive the world had ever known, disappeared in rapid stages in the 20-odd years after 1945. This process of decline, by which Britain lost the world-power status it had held for nearly 200 years previously, became a core interest for many people: for intellectuals in general, but also people of specific political persuasions. In the political sphere, the demise was a source of regret generally for those on the Right, and of satisfaction generally for those on the Left.

The issue of loss of status has remained a live one. Many on the Right, and some in the Centre, having lamented the loss for many decades, have welcomed the recent resurgence of militancy in British foreign policy under the New Labour government. The complete opposite is the case for almost all those on the Left.

Regarding the advent of the Welfare State: this has been viewed by many people, especially those on the moderate Left, as the definitive feature of life in British society since

1945. The National Health Service; the benefits system in connection with unemployment, income support, disability, and other areas where people without substantial private means need help: these state-provided facilities are seen as marking the most significant of all departures from pre-war society, in which state-provision was minimal.

The above use of the phrase 'the moderate Left' needs emphasis. In contrast to the moderates, those on the far Left, chiefly Marxist-Leninists, have seen the Welfare State as only a small step in the right direction, as merely an aspect of what they regard as reformist capitalism—a capitalism attempting to adjust itself to general social needs in order to consolidate and perpetuate itself. The far Left seek the complete overthrow of the capitalist system, and so reject any reformed version of it. This has been their position since (and before) 1945, and is the more emphatically so now (2009) given the present economic crisis, defined by them as the latest inevitable crisis of capitalism.

However, in contrast to the above attitude, it is clear that the Welfare State idea is, at least in part, the outcome of general humanitarian and socially radical schools of thought, dating back several centuries, to periods long before modern capitalism came into being. Certain tendencies in mediaeval thought, and especially in the revolutionary period in the 17th century, are among the idea's forerunners.

At the same time, there is no question of the overall impact of capitalism on modern thinking, in connection with the Welfare State and several other things besides; and for many people on the Left, moderate as well as extremist, the workings of the capitalist system have been the core intellectual interest of the post-war period: the issue in terms of which most other issues have been viewed—society's pivotal feature.

That capitalism is a factor of leading importance no informed person could possibly deny. However, that it surpasses all other issues in significance is less certain. Since

1945, there have been a number of developments in British society which are of major import even though they are either only partly related to capitalism, or not related at all.

In dealing with these issues, I will, where appropriate, comment on their connection or non-connection with capitalism. This is a necessary procedure for conveying the enormous complexity of social phenomena over the last 60-odd years, a complexity which often goes well beyond the economic sphere.

The first of these phenomena is the decline of religious belief: this relates Britain, in varying degree, to a good deal of the rest of the Western world. The factors bound up with this decline are of course multiple and complex. Also, many were operative well before 1945, but have gained in momentum since then. They include: the cumulative impact of scientific/empirical thinking; the demise in social influence of ecclesiastical institutions; the lingering shocks of war and the upheavals war brings.

It should be added that, in these and subsequent observations, account is being taken of the fact that there has been a resurgence of religious belief in Britain. However, this phenomenon, over the past decade or two, does not invalidate the argument that the last six decades have seen, on balance, a decline in belief.

Taking the first of these factors, the growth of the scientific outlook: while it is true that the advancement of science, especially technology, has partly been connected with capitalist expansion, much of it, particularly in the pure sciences, has had a momentum of its own. It perhaps goes without saying that a good deal of pure science investigates areas of reality which lie outside of, and beyond the possibility of exploitation by, the workings of the economic system. Further, concerning the linkage that does exist between the growth of science and that of capitalism, there is no automatic connection, in the West, between the enlargement of capitalism and the decline of religious belief. Historically, the origins of Western capitalism are closely linked to a religious outlook—Christian Protestantism; and many capitalists remain religious believers, whether Christian Protestant or otherwise.

Taking the second factor, the waning influence of ecclesiastical institutions: this has been especially marked in the reduction of the communal role which the churches have traditionally played in British society.[1] Generally, the diminution of influence has been the effect of belief-decline, though there are some other causes, economic and political, as well.

Regarding the third factor, the effects of war: this was more an issue in the period immediately after 1945 than it has been in recent decades, and for obvious reasons. Mainly in the popular mind, the stupendous havoc and destruction wrought by World War Two was regarded as a challenge to the religious doctrine, associated chiefly with Christianity, that there exists a deity who is not only all-loving but also all-powerful, and therefore able to prevent certain human beings from exercising their free will to commit atrocities which bring untold suffering to millions of others. For a number of people retaining a memory of the war, this argument has remained a cogent one.

As regards the roots of WW2, there is no doubt that capitalism was a causal factor. However, there were other important causes as well: nationalistic, racist, and ideological ones—not themselves the products of capitalism but of other factors, of historical and cultural kinds.[2]

---

[1]   A role, incidentally, which one of Britain's leading post-war poets, Philip Larkin, looks back on as a non-believer in his famous 1955 poem, 'Church Going'.

[2]   A leading cause of World War Two, Nazism, was a nationalistic-racist ideology which had complex cultural origins and which was not an outgrowth of German capitalism in the post-WW1 period. This is the case even though it was initially financed by, among others, leading capitalists. The point that it was not a product of capitalism is also compatible with the fact that its advent was connected with the economic crisis in Germany in the 1920s—a crisis caused largely by the financial system associated with capitalism. Economic crisis gave Nazism its opportunity, but not its doctrine. A further point is that

Overall, the psychological effects of the war have played a leading role in the erosion of religious belief. This role has been at least as extensive as that of the psychological effects of WW1. Indeed, it may have been more so, given the advent of nuclear weapons in the closing part of WW2, and the accompanying realisation, by a large number of people, that the natural world contains properties which are colossally dangerous. Such a realisation is, in the eyes of many, incompatible with the argument that the natural world is the creation of a benign deity.

In addition to the several causes of belief-decline, one of the latter's chief results, at least among people of strong philosophical tendency, has been to produce a sense of cosmic meaninglessness: a view that the cosmos is both dangerous and without moral significance. A painfully changed picture of the universe partly accounts for the appeal, among discerning readers of philosophy and literature in Britain, of the writings of Sartre and Beckett. Sartre's influence in this country dates from the late 1940s, and Beckett's from the mid-50s. The impact of these two authors, and of others of similar outlook, has been widely and lastingly felt.

To some extent linked with an altered cosmic outlook has been the extensive questioning of a number of thought-patterns traditionally associated with religious culture. Many of the latter are to do with social mores and arrangements. Partly as a result of the questioning and critical attitude, Britain has seen, for example, a decline in the institution of the extended family: a demise which has been a prominent feature of social life mainly since the 1960s (much less so in the period from 1945–60). Indeed, with the diminution of the extended family has come even that of the nuclear family. This has been evident in the growth of one-

---

neither Nazism, nor anything like it, had emerged in the 40-odd years of German capitalism prior to WW1.

Likewise, Italian fascism, another major cause of WW2, was not an outgrowth of Italian capitalism, but had complex historical and cultural roots.

parent families (the parent being of either gender) and in the increasing divorce rate. Much has been written about this process of atomisation, and the literature on it is of a complexity and extent unparalleled in British social studies.

Also, it should be noted that atomisation of this kind has only some links with the operations of capitalism. One link is the technological mobility of capitalism, which affects the employment situation—which in turn affects family cohesion. Yet these links are limited; in other societies with a capitalist system, such as France, Italy, and Japan, family unity remains generally stronger than in Britain. This is due to broad cultural factors which are different from those obtaining in this country.[3]

Atomisation in Britain has, to some extent, been connected with the expansion of Welfare-State provision. This may at first seem an odd statement, given the fact that the State provides a range of family benefits. But the extent of its provision also to individuals has meant that the latter can be more economically independent of the family unit than was possible in pre-Welfare State days. Again, comparing the situation with that of certain other countries, there is less atomisation in France and Italy, which also have Welfare State provision. Once more, the discrepancy is mainly due to the cultural factors previously referred to.

Aside from the issue of atomisation, the Welfare State can be seen as productive of immense social variety. There is of course some relation between atomisation and variety, since variegation always, to some degree, produces problems; but here we are focusing on the positive aspects of diversity. The Welfare State, in seeking to create equality of educational opportunity, has helped to establish genuine meritocracy in British society on a scale unimaginable before World War Two. This is not to say that no more needs to be done. Clearly, with the current (2009) cutbacks in public spending,

---

3   Also, in the case of France and Italy, the cultural factors include religious influence, which remains stronger than in Britain. The influence in question is that of Roman Catholicism.

especially in education, there are real dangers that the expansion of meritocracy will be—indeed, is being—seriously impeded. The cutbacks must be reversed.

In highlighting meritocratic values in relation to the Welfare State, we again have to consider the latter's connection to capitalism. No doubt capitalists have viewed State-aided meritocracy in terms of increasing the mastery of skills which can then be inputted into the industrial, financial, and commercial systems. But the consequences of meritocracy have extended much beyond such purposes. They have produced a social and cultural situation so multi-faceted that it cannot begin to be adequately understood just in relation to the capitalist system.

The meritocratic principle must be seen as a leading feature of British social thought sine 1945 (incidentally, as an object of both endorsement and non-endorsement). Viewed positively, the principle has been regarded as bracing and challenging, entailing stringent appraisal of individual achievement, especially in the arts, the sciences, and areas of social contribution: the kinds of achievement which go beyond the rigid 'class' categories which still vitiate much thinking in certain sections of society, including the Marxist Left. At the heart of the meritocratic argument is a concentration on issues such as the range and calibre of the individual's consciousness, his/her powers of criticism and discrimination, and the quality of his/her moral energy.[4]

If Britain remains the open society which, to a considerable extent, it now is, then this evaluative rigour will continue; and such will be the case even if the capitalist system is replaced, largely or wholly, by a socialist one.

---

[4]   These issues renew interest in certain kinds of 19th century British thinking on the subject of exceptional individuality; for example, that of John Stuart Mill and Matthew Arnold. In fact, the modern concept of meritocracy bears a striking resemblance to Arnold's notion of 'the remnant': a group of people whose education and outlook are so informed and advanced that their thinking completely transcends 'class' attitudes.

Further, affirmation of meritocratic values by no means excludes considerations of the problems, previously examined, to do with the decline of religious belief. As indicated, meritocracy is about quality of total outlook as well as that of particular activities. The true meritocrat is the person who has intellectually stretched himself/herself to the fullest possible extent: who has experienced to the maximum the complexities of modernity which Popper describes as "the strain of civilisation"; and who has made maximal effort to cover all the chief areas of modern knowledge which Malraux characterises as "la musee imaginaire". S/he is even the person whose mental and emotional experiences open the way to sympathetic understanding of Sartre's dictum that "life begins on the other side of despair".

The above references to three thinkers who are not British carry the clear implication that the foundation of a merit-ocratic and intellectually open society transcends national boundaries. Such a foundation exists in British society, and the hope is that it will be continually reinforced.

Now for a few comments in extension of some of the points made previously. To what has been said about the decline of the family, it must be added that sexual mores have changed considerably in Britain—especially, again, since the 1960s. There has been a major increase in pre-marital sex among teenagers and young adults, and in extra-marital sex (the latter obviously linked with the afore-mentioned increase in divorce rates). This is in line with tendencies in certain other Western countries, especially the United States. These tendencies have been collectively described as 'the sexual revolution'.

As with the demise of the family unit, this development has been the subject of an extensive literature, again unprecedented in British social studies. Explaining it involves reference, once more, to many factors. These again include the decline of religious belief and of thought-patterns traditionally associated with religious culture. Such decline has much lessened the stigma traditionally connected with non-marital pregnancy. Also, there have been improved methods of contraception, as a result of advances in medical technology, plus the increase in availability of such methods. Further, abortion facilities have amplified.

In connection with developments in the fields of contraception and abortion, the Welfare State has again played a major role, through the National Health Service. The NHS has also figured prominently in the treatment of venereal diseases.

Bound up with the sexual revolution are economic factors as well: the increasing economic independence of women, especially career women who opt for sexual relations without marriage; and the decreasing legal costs of divorce.

Capitalism figures too in various ways, especially in offering top jobs to more and more career women, so enhancing their economic self-sufficiency. However, the role of the economic system must again be seen in conjunction with a host of non-economic causes and conditions. Some people will argue that capitalism is actually the central consideration in the changing of sexual mores since, without the taxes drawn by the government from a capitalism which flourished for a long period after WW2, the NHS would never have been able to provide the various facilities which support this change. This point is certainly an important one, but it still must be seen in context. The extensive channelling of State resources into areas such as contraception, abortion, and disease-treatment has resulted not only from available wealth but also from *public moral attitudes.* The latter, as previously said, have altered immensely in the post-war period. Without the change in moral outlook, it is far from certain that economic prosperity alone would have led to these developments in public health practice. Hence, available wealth must be seen as only a necessary condition, not a sufficient one, for radical alteration in public policy.

There is no automatic relation between the capitalist system and the modern sexual outlook in Britain. A study of the Victorian era, for example, shows that during this period capitalism was at high tide, and yet there prevailed, as an official and public moral attitude, an endorsement of a rigid code of sexual behaviour. True, capitalism and State activity were much more separate than in modern times, but there is nothing to show that, had they been more connected, sexual permissiveness of the modern kind would have been endorsed. An examination of cultural and social factors in the Victorian period strongly suggests that they would not. As a related point, we should remember that one of the roots of capitalism in the West was religious, with many early capitalists coming from a culture which they saw as one of high moral rectitude.

Moving on now to other aspects of change in sexual attitude: there has been an increasing public acceptance of homosexuality, in both its male and female forms. It is worth remembering that male homosexuality was not legalised in Britain until 1967, almost a generation after the end of World War Two and the establishment of the Welfare State. Up until then, the fact of homosexuality had, in most hetero-sexual circles, been either close to unmentionable or the object of jokes. However, since legalisation we have seen a growth, at all social levels, in mature and sympathetic consideration of the subject, as well as much more openness, especially in the mass media, in unbiased reference. A parallel development is to be seen in some other Western countries, particularly the United States. Partly as a result of amplified acceptance, homosexuals themselves have become increasingly confident and expressive in the public arena. Finally, there appears to be little if any connection between these developments and the workings of capitalism.

Not actually part of, but in connection with, alteration in sexual outlook has been the expansion of feminism. Manifestly, links exist between this and the growing economic independence of women previously mentioned.

To the extent that capitalism has played a role in this growth, it has contributed to the expansion. But again roles have been played by other things too: the enlargement of gender equality in educational opportunity, and the expansion of meritocratic appraisal of people as individuals, regardless of gender.

Leaving the area of sexual attitudes, and returning to the political sphere with which the essay began: in terms of constructive and progressive political protest since 1945, praise must go, almost invariably, to the broad Left. This point links with one made in the opening part of the essay: that the Left has opposed recent militarism in British foreign policy. Over the years, it has protested against militarism as a function of imperialism, and as pursued by the British political establishment and by its closest ally, the U.S. political establishment. In this regard, we should note that, for over six decades, very few large-scale popular demonstrations against an aspect of foreign policy have been organised by the political Right, whereas the Left has organised a very large number indeed.

While it is true, as said, that most of the British Empire disintegrated rapidly in the 20-odd years after World War Two, in the first decade of that period there was much imperialist violence directed against independence movements, in such countries as Burma, Malaya, Kenya, and British Guiana. There was also action in Egypt, connected with the Suez Canal. Further, in subsequent years, in the few remaining outposts of empire, violence continued. More recently, protests against foreign policy have been prompted by that policy's complicity with the U.S. incursions into Iraq and Afghanistan; and this latest phase of protest continues.

The issue of imperialism—both British and American—clearly relates to capitalism; so, in this sphere, reference to the economic system is absolutely central. British action in, for example, Malaya, British Guiana, and Kenya was bound up with these countries' economic resources and their utilisation in the global capitalist system. Similarly, British action in Suez was motivated by shipping and commercial interests. American action in Vietnam (for which British governmental support was mainly political rather than military) was partly actuated by consideration of South East Asia's mineral resources. (It should be mentioned in passing that protest against American incursion into Vietnam and other parts of Indo-China was a focus for many prog-ressively minded people in Britain in the late 1960s and early 70s, as it was in other Western European countries such as France and West Germany.) Joint American and British action in Iraq is unquestionably linked to that country's immense oil reserves. The same joint action in Afghanistan has geo-strategic implications rather than economic ones, though it must also be noted that part of this geo-strategic importance is a pacified Afghanistan's possible utility as a conduit for oil from the enormous Caspian Basin reserves to the Indian Ocean.

Given the crucial role played by capitalist interests in the formation of foreign policy, it is small wonder that protest has come largely from anti-capitalists, both those who are opposed to all forms of capitalism and those who oppose only monopoly capitalism: these being the two main groups comprising the broad Left.

Further, the broad Left is socially complex, and includes, for instance, manual workers, people in technical and administrative positions mainly in the public services, and journalists and teachers. If one is concerned with 'class' categories, then this group is extremely hard to classify. Its eclectic character perhaps reflects the greater fluidity in British society since 1945: the greater equality of educational opportunity and cultural access, and the freer mingling of different social groups. This eclecticism contrasts with the generally simpler social make-up of many Leftist protest groups before WW2.

In conclusion, let us focus on the key protest issue of the last 10 years: U.S. British incursion into Iraq and Afghan-istan. Though the broad Left have been the main protesters, it must in fairness be added that this issue has exercised a wide range of democratic and progressively-minded people in Britain. The concern of all has been for the welfare and betterment of Islamic societies, in the face of what is seen as Western aggression and interference.

However, for those many British people who are agnostic or atheistic, there is a problem connected with the fact that staunch religious belief is one of the most prominent features of the societies they wish to help. The question has arisen: in order to give help, should concessions, even extensive ones, be made to the religious mentality, or should the religious factor go completely unmentioned? Some progressives have decided to keep silent on this issue, others not. The problem is a highly complex one, and is ongoing. However, difficult though it is, it is an aspect of engagement with a pivotal world issue, and praise is due to British progressives for taking on this engagement.

Liberal Politics &
Literature

The relation between politics and literature is part of a larger relation between politics and culture as a whole, literature being of course a component of the latter.

Culture can be defined as a entire way of life, a richly informal integration of many different kinds of experience. By contrast, politics is a highly formal set of activities, and not aiming to encompass the full range of available experience. Its activities are highly institutionalised, and subject to impersonal, protocol procedures. However, *liberal* politics, while still working within these restrictions, seeks, more than any other kind does, to relate itself to the broader sphere of experience. To an exceptional degree, it tries to promote a general state of social flourishing which renders a vibrant culture possible. Its objective is to support a wider way of living and interacting, one far less dependent on formalities and far more oriented to the personal, the spontaneous, and the creative.

However, even the most liberal kind of politics cannot escape its institutional limitations. Its basic way of thinking is one it shares with all politics: it is, as Aldous Huxley says, "the science of averages"; and, as Auden states, it is "concerned with large numbers of people, hence with the human average". Thus the political mind does not, and, being political, actually *cannot,* extend sufficiently into the sphere of the personal, unique, eccentric, and private, the unpredictably variegated and complex, the idiosyncratic and impromptu: in other words, into the sphere of culture. It can assist this sphere, but never become it, or be mentally co-extensive with it. Awareness of this point should always attend efforts at political reform and renewal (efforts which are in themselves, of course, perfectly valid).

Given the broad definitions of culture stated above, let us now turn to literature. Arguably, creative writing constitutes the most expansive mode of imaginative expression available in any culture. This is really to say that the word is the most capacious vehicle for imaginative expression yet devised. A large claim, yes, but it is one which many people will endorse. In fact, when one thinks of the role played by the word not only in literature itself (fiction, drama, and poetry) but also in extensively visual art forms such as cinema and opera, one may regard the claim as perhaps not so large after all.

It has often been said that, between them, the novel, the play, and the poem (epic, dramatic, lyrical) have provided us with a greater sense of human variety, in terms of personality, experience, and situation, than has any other product of the human mind. For the modern liberal mentality, diversity is of course a pivotal concern and value. At the same time, it is important to emphasise that such variegation distinguishes the great literature of all periods, including those which pre-date modern liberalism. One has only to think of Shakespeare and Chaucer, who presented massive variety to the readers of their day, but within a cultural framework which differed markedly from the modern liberal one.

Liberal politics can actually do no more than operate, within its own framework, in such a way as to acknowledge mankind's diversity as perennial, and to attune activities to that acknowledgment. Further, the variety it should recognise covers not only types of people but types of situation. The latter include the deeply problematic, the intractable, and even the tragic. In fact, the existence of such kinds of situation accounts for the production of certain works of literature which have been created as the only type of alleviative response to problems encountered. This means that practical solutions to difficulties, solutions of the sort that political and social agencies can offer, have been seen to be unavailable; and instead a way of enduring the problem, the only way of enduring it, has been found in art. That literature may, in certain circumstances, be this only way is a fact which the political and social spheres should recognise.

It must be added that the existence of insoluble problems is something which liberal politics, indeed the liberal mind in general, frequently fails to see (this perception is, by contrast, quite often found in the conservative mentality). A clearer outlook would better enable liberal politics to combine, on the one hand, good intentions and well-meaning solicitude with, on the other, unerring realism: a realism which would tell it how far it could effectively go in offering help to the writer, and where that help must stop.

Again in connection with the boundaries of liberal political action, it should be remembered that much major literature has arisen from conditions of long-standing social and cultural stability: a stability to which political systems, whether liberal or otherwise, have only partly contributed — indeed, *can* only partly contribute. A social and cultural order which is complex and historically important is always much more than the political system connected with it; it can therefore never be fully explained in terms of that system. It is not from the latter but from the settled, or relatively settled, social and cultural situation that most of the materials of much great literature are drawn. Examples, taken from all periods of history, include: Virgil's *Aeneid,* Dante's *Divine Comedy*, Chaucer's *Canterbury Tales,* Tolstoy's *Anna Karenina,* George Eliot's *Middlemarch,* and Proust's *Remembrance of Things Past.*

This of course is not to say that writers completely endorse the contexts on which they depend for their material. Such total acceptance is rare, for obvious moral reasons. But it is the case that their work has a tremendous density of detail because they fully know their way around the social situations from which they draw their material; and the garnering of such knowledge is a gradual process, requiring comparative stability. Again, that knowledge is of much more than the political system. (It is even of much more than the economic system.)s Indeed, in the texts specified above, and in many other examples that could be given, details on political systems figure hardly at all.

Major writers always stand in a complex relationship to a complex environment. They possess a strong sense of its past as well as of its present, and a keen awareness of its moral issues and problems. This multi-faceted view may not always be encompassed by liberal politics because the latter's frequent assumption is that more difficulties are capable of being resolved than is actually the case. The best features of liberal politics—its solicitude, generosity, and authentic concern for social flourishing—remain perennially valid. But, as we have seen, even these approaches have their limits when it comes to the intricate and many-layered areas of experience with which major literature deals. Bound up with this is the fact that, to repeat, they can never, as *political* approaches, adequately encompass the individual context.

The Idea of
a Classless Society

This idea, in its modern form, originates with socialist schools of thought, especially Marxism. Socialist thinking usually deploys the word 'classless' to mean 'economically undifferentiated': a phrase denoting the view that society should not contain economic divisions, at least not ones large enough to constitute 'class' differences. The view is that everybody should be on the same level, or on very similar levels, economically.

This position has a good deal to be said for it. In an economically undifferentiated society, no one individual or group could wield material power over others; or enjoy economic advantage in waging political campaigns; or have special influence over the body politic; or thrive economically while others suffered. There would not be, then, at least four major problems with which world society is currently beset. Also, if incomes were the same or similar, motives for taking on work which was very challenging, complex, and socially important would probably be wholly moral in character, since the incentive of so-called 'competitive' salaries would be absent. Thus there would exist more likelihood of the better kind of person choosing such work, despite the extensive training that might be required for it.

Other points could also be made in favour of the idea of classlessness, but we must now examine its limitations — specifically, those areas of social fact which it does not cover and has nothing to say about. This is not to argue that the advantages which its implementation would bring are not real advantages; as the previous paragraph has indicated, they unquestionably are. But what it is to contend is that the realisation of these advantages would still leave a range of problems: problems which certainly obtain in societies that are economically divided but which, arguably, are not confined to such societies.

These difficulties are bound up with the diversity of what can be called natural ability. If one accepts the view that capacity has pre-natal sources, largely hereditary ones, and not environmental ones, then in this sense one can use the phrase 'natural ability' and go on to speak of the enormous variegation of these abilities across society as a whole. At this point, advocates of the classless society may well reply that there is no conflict between economic equality and appreciation of ability-diversity; and, yes, ideally, there should not be. But in practice this may not turn out to be the case. It must be noted that diversity in capacity is a matter not only of variety but also, at least in an open society where appraisal can be freely expressed, of *gradation*. Where the terms 'excellent', 'good', 'average', and 'below average' are openly used, the concept of ability-hierarchy is operative, and this concept plays a large part in determining the individual's place in social and cultural contexts.

Such is already the case in economically divided societies, with regard to a wide range of capacities — artistic, scientific, philosophical — many of which have little or nothing to do with the current economic system; and, in any open society of the future, such will continue to be the case, even in the event of the transformation of the economic system.

Now it is a hard fact that, in present-day society, the notion of ability-hierarchy produces tension, even hostility. Envy and dislike are often felt by those of lesser ability toward those of greater. Further, protective egotism frequently comes into play as a way of internally resisting recognition of superior capacity. In fact, all kinds of tortuous psychological processes are activated by the encounter with higher ability. It is true that some forms of resentment are, in present society, caused by differences in pay and economic status; and these forms would not exist in an economically standardised society. But even in the latter, other kinds of resentment might well continue. Unless transformations in the economic system can produce transformations in certain basic psychological traits—ones which great minds across the centuries, and across a range of different socio-economic contexts, have accurately described—then the persistence of these traits is a distinct possibility in any future form of society.

The above range of problems is obviously connected with the issue of meritocracy: an issue of great importance in the modern world, especially in Western society. Some writers have argued that differences in natural ability should not be the basis for social differentiation, that social differentiation *of any kind* is a bad thing, because it under-mines a sense of community and social solidarity. There is substance in this argument.[1] At the same time, it is hard to see how any complex society like our own can avoid, either in specific occupational or general cultural terms, making appraisals of ability, and appraisals which inescapably lead to social practices and policies. As long as society remains as occupationally and culturally complex as it is now, a sense of differentiation must surely continue, even if that different-iation is no longer of the kind which has traditionally come under the heading of 'class'.

This last point needs further attention. The traditional concept of class has had mainly economic content: has been chiefly to do with economic status (while also possessing certain cultural implications, e.g. that higher economic status means more opportunity for education and culture). In a society which was without large economic divisions but which was also meritocratic, the word 'class' could be used

---

[1]   An argument which is articulated with notable clarity by Raymond Williams in *Culture and Society 1780–1950* (1958), especially in the chapter on T.S. Eliot.

in new ways, to mean ability-groups.[2] In such a society, appraisal would clearly be operative, with capacity-estimates being made as a matter of course. In this regard, it would resemble present-day society, in so far as the latter does make genuinely meritocratic judgments.

---

[2]   This line of thinking, which equates social definition with both type and level of ability, has been explored by, among others, the sociologist Karl Mannheim.

Questions Facing Socialism as a Cultural Outlook

By 'socialism as a cultural outlook' is meant socialism as *more than* the advocacy or practice of an economic programme: that of public ownership and control of the means of material production, and of the methods by which material products are distributed throughout society. The additional meaning is advocacy or practice of a particular kind of cultural life. A socialist vision of culture must, virtually by definition, be an extremely democratic one, in which there is the maximal possible sharing of ideas, emotions, and perspectives: the largest conceivable amount of common psychological currency. This implies the highest possible degree of what can be called equality of cultural condition.

This type of equality is not to be confused with the concept of equality of opportunity for general cultural experience, which emphatically does not imply a post-opportunity equality of condition.

On the overall question of cultural condition, one could, in passing, point out that the question is especially important in secular society, where no pre-established mindset is given to people to imbibe in a largely unquestioning fashion, unlike what usually happens in most religious cultures. In the secular context, the attaining of a mindset that has been fully thought out by the individual concerned, and can therefore be defended by him/her on purely empirical, rational, and logical grounds, is a definite cultural achievement: something *made* and not given.

As regards distinctive cultural achievements in general, whatever the context, the question arises as to how such attainments, which require hard, prolonged, and sometimes solitary effort, can be squared with the socialist concept of a general cultural sharing. It is surely the case that the attainments can be fully shared *only* with those people who have themselves made endeavours of the same or similar kind; and shared to a lesser but still significant extent with those who have not themselves made the effort but have a strong critical appreciation of the latter. That, obviously, will not be everyone. Endeavour is recognised by fellow-endeavour, and by its attentive observers; like is known by like.

For example, in philosophy, the detailed construction of a viable philosophical outlook is no easy matter, and neither is the adequate comprehension of it by others. The same is true of the creation of a major work of art (in any medium) and of advances in the sciences. So, where there is inequality in effort or in appreciation of effort, there must also be inequality in cultural condition.

In response to this view, many socialists argue that the establishing of an egalitarian economic system will have such a liberating effect on human potential that the cultural efforts of all will increase to an extent unparalleled in any previous period of mankind's history. That only a few have so far achieved extraordinary things is mainly due, they contend, to cultural deprivation resulting from inegalitarian economic systems. With the ending of such systems and the deprivation which goes with them, cultural output will amplify to such a degree, not only in quantity but also in quality, that the traditional division between exception and average, the few and the many, will disappear. The kinds of endeavour and appreciation of endeavour which, up to now, have been unusual will become the norm.

The above argument, it need hardly be added, is premised on the assumption that everybody except the educationally subnormal has roughly the same potential, and that the same was true of the past. Hence past differences in cultural performance have chiefly stemmed from factors other than the possession of capacity, as do present differences.

The aforementioned assumption is certainly worth considering, given the fact that it is impossible for us to investigate the total amount of potential that has existed in past societies. Hence our lack of knowledge about the past means that the assumption cannot be refuted. However, that same lack of knowledge also means that the assumption cannot be vindicated either. So, it can only have the status of a surmise, one not subject to empirical verification or falsification; it can generate no fruitful research programmes; simply, nothing can be done with it.

Further, though it cannot be disproven, it can at least be looked at in the light of present-day experience: an area which definitely provides empirical support for postulates about culture and society. If we take, for example, British society since 1945, we see that the very real expansion in educational and cultural opportunity evident in the last 60-odd years has certainly produced an increase in the *quantity* of cultural production, especially in the kinds of products emanating from the mass-media. Nevertheless, it is arguable that there has not been a comparable increase in the *quality* of production. The best that has been produced in this period—many discerning people would claim—is not superior to the best produced in past eras, when the extent of educational and cultural opportunity was much narrower. Moreover, this modern best has been the work of a few, as was the case with the best in the past. A further point is that the continuing status of traditional 'classic' works in modern culture, especially those in music, literature, and philosophy, signifies recognition that present and recent attainments do not constitute a qualitative advance over those of the past.

Also, if we look at the privileged social groups which, not only since WW2 but long before it, have enjoyed the widest educational and cultural advantages, we find that they have this in common with the groups who have only enjoyed such advantages in post-war society: only a minority among them have achieved cultural distinction.

The above considerations tend to suggest the following: that it is not true that everybody who is not educationally subnormal possesses roughly the same potential; that the capacity to achieve things traditionally regarded as extra-ordinary is, by its nature, rare; and that, therefore, the attaining of those things will remain the exception.

This reasoning chimes not only with six decades of social experience but also with a viewpoint voiced across many centuries by a number of eminent minds, some as far apart historically as Plato and Schopenhauer. It must be concluded that this viewpoint also, in the past, arose empirically, from solid social experience. The argument therefore carries weight. The serious doubt it casts on the socialist assump-tions previously examined finds reinforcement when—just to confine ourselves to literature—we consider the calibre of works such as: Dante's *Divine Comedy,* Shakespeare's *Hamlet* and *King Lear,* Cervantes' *Don Quixote,* Milton's *Paradise Lost,* Goethe's *Faust,* Tolstoy's *War and Peace,* Dostoyevsky's *The Brothers Karamazov,* Melville's *Moby Dick,* and Proust's *Remembrance of Things Past.* When, in connection with these works and others of comparable quality, we ask ourselves if the ability displayed, either in writing them or extensively appreciating them, is and always has been the possession of the majority, we are likely to return a negative reply.

Further questions which must be directed towards socialists are as follows. Regarding the many individuals in society who appear to be relatively content doing jobs which place no extensive demand on the intellect: is this content-ment mainly or entirely the result of educational or cultural deprivation? If more educational and cultural stimuli had been provided, would that contentment never have taken shape, or would it evaporate if, now, such stimuli were provided? Next, regarding the many individuals (a number of them being in the previous group) who show themselves to be indifferent to—even hostile to—the kind of mental effort required for advanced intellectual culture: are these attitudes mainly or entirely the product of the afore-mentioned deprivation? Again, would they never have come into being, had early environment been different, and would they now vanish if previous deprivation were made good? Finally, can the same questions be applied to all those people, including the ones already mentioned, who opt for economic or psychological security, or both, rather than a life of intellectual or artistic adventure? Unless socialists feel able to return an *absolutely certain* 'yes' reply to each of these questions, they will need to reconsider, in depth, a number of their presuppositions.

It is important to insist on these considerations because failure to do so blurs the distinction which must be drawn between the concept of a socialist culture and that of a meritocratic one. The latter does not contain the notion of equality of cultural condition, only that of equality of cultural opportunity. The two concepts should not be conflated, though unfortunately this is done by some socialists and some other people as well.

Such conflation involves equating liberty with equality: two things which, though closely related, are categorically separate. The distinction between them becomes completely clear when we relate both to the idea of opportunity. Equality of opportunity opens the way for freedom to develop in all different kinds of ways, and to different extents. In this sense, difference emanates from similarity, and diversity from an initial standardisation.

The crucial importance of diversity will be evident to anyone with an extensive knowledge of cultural history. For example, in the arts: the enormous range of media, of genres within media, and of styles within those genres, are a triumph of variety and complexity, and an inexhaustible source of fascination. Likewise in philosophy: the vast spectrum of viewpoints, the different modes of argument-ation, and the nuances of difference even among positions which broadly concur with each other, are again a marvel of plurality. The same can be said for manifold achievements in the sciences.

Scintillating variegation and outstanding individuality are surely their own justification. Yet it is unclear if a socialist view of culture truly values them. Its collectivist emphasis, and stress on commonality, implies that it is less interested in individual distinction than in group cohesion—and in the latter, indeed, on a large scale. The argument that a super-capable few exists amidst a majority of only average capacity is not one socialism explores in detail, essentially because of its refusal, or at least reluctance, to accept the idea of the generically exceptional.

Hence, among a number of cultural considerations, socialism is not much interested in the concept of grandeur: understood as that which soars high above averageness, and does so without embarrassment and with a good conscience. Yet all major cultures known to history have had this concept, and have acted on it. While socialists, and also other extreme democrats, may well reply by citing the undeniable fact that all past societies have been based on gross economic inequality, this observation does not actually invalidate the notion of grandeur; if grandeur could exist in a context free of the injustices of the past, there would be no moral objection to it as such. Even as matters stand, many people admire existing and surviving examples of grandeur, especially in architecture, epic poetry, and music, despite knowing that they were produced in the context of inegalitarian social conditions. There is a strong case for arguing that a culture must always afford space for that which radically elevates, or offers to elevate, human horizons; and liberty to utilise that space is a fundamental and indispensable liberty.

This essay began by drawing a distinction between socialism as an economic programme and socialism as a cultural outlook. Focusing on the latter, we have said little on the former. Suffice it now to advance the argument that a form of socialism, defined as advocating the public ownership and control of all *major* processes of material production and distribution, is the most progressive approach to the present-day world's economic problems. It is the one most likely to minimise social and economic injustice, and to achieve long-term economic stability. Further, it is the most probable guarantor of equality in civil and human rights, and in social opportunity. And more: its general adoption would be one of the most decisive turning-points in the world's economic history.

However, alongside this large endorsement, a reminder is perhaps needed that there remain the cultural perspectives advocated in this essay, ones critical of socialist views on culture. Clearly, acceptance of a certain kind of socialist position in economics does not entail acceptance of socialist positions in other spheres as well. There is nothing wrong with only partial acceptance of a particular outlook. In a staggeringly complex world—the complexity characterising the past as much as the present—no single doctrine or way of thinking can possibly be all-comprehending, or should claim to be. A realistic outlook results from close observation of a very sustained kind, and the more sustained it is, the more synthetic and flexible the outlook will be. These latter qualities are essential for an adequate response to the multi-facetedness of any advanced culture, because such a response incorporates an increasing range of perspectives.[1] In the cultural sphere, a finely tuned and carefully modulated mode of apprehension is called for; and this is one which the socialist outlook seems not to possess sufficiently.

In terms of economic issues, the modern world is indubitably at a crossroads. There is no doubt that monopoly capitalism, especially in the sphere of finance capital, is in massive crisis. Radically new economic arrangements are needed; and this essay has briefly argued what they should be. Nevertheless, while arrangements and systems should be instigated in the economic field, the same should not be

---

[1]   In these respects, incidentally, the shortcomings of the short-lived artistic doctrine of socialist realism in the former Soviet Union will be obvious.

attempted in the cultural field. The latter ought to be viewed as a completely open area, to be left to itself to evolve in whatever ways it may. Anyone genuinely concerned about culture ought to be of the view that the most that can be done, in the way of economic and political effort, is to establish the social conditions which enable equal access to educational and cultural opportunity. That opportunity is the doorway to the rich multifariousness we have been discussing. As regards future types of culture, no preconceptions or prescriptions should be entertained, no cultural formulae advocated — but with one vital exception: that cultural achievements reflect as much as possible the enormous magnitude of human reality and experience. In doing so, they will link with the great attainments of the past.

The Welfare State Liberal

This essay reflects a number of points made in the previous four. The Welfare State has been discussed, if fairly briefly, and so has liberalism as a political outlook relating to cultural activity. Now, more needs to be said on the liberal mentality in general. Its essence, I would argue, is openness: to information and knowledge, to differing attitudes and values, and to disparities in social and cultural context. Hence, one of its key activities is viewing the enormous variety of social phenomena in the light of each other: weighing discrepant things against each other from the standpoint of genuine interest and responsiveness. 'One of its key activities', to repeat: for the liberal, like everyone else, must of course do more than be interested and responsive. He must, at some point, come to conclusions, must affirm and assert, both morally and culturally. However, the path leading to his eventual assertions is longer, more winding, more strewn with the multi-coloured flowers of awareness, than is that of the non-liberal.

Clearly, the taking of that path depends in the main on educational opportunity; and here the Welfare State, as it has existed since 1945, has played a crucial role in providing public education to an extent never previously available. At the same time, the point must be made that the public education system has stimulated people to embark on the liberal journey *only* to the extent that it has *not* imposed any kind of cultural closure: only in so far as it has opened up cultural vistas, not occluded them. To the degree that it has committed itself to cultural openness, it has refused to present its materials through the distorting lens of an outlook based on categories of class, nation, ethnicity, and (so-called) race. Its finest product, then, has been the mind free of prejudices of the class, national, ethnic, and racist kinds: that is, the truly liberal mind.

Reference was made in an earlier essay to the 19th century concept, framed by Matthew Arnold, of 'the remnant': those people who, through education, have moved completely beyond 'class' mentalities. By implication, the concept can be taken to mean moving beyond sectarian outlooks of *all* kinds. No pre-1945 social idea is more pivotal than this one to the notion of the Welfare State liberal, the person aspiring to universalist vision, and receptive to all that the human adventure has forged in time and place.

Unfortunately, in opposition to this outlook, and over six decades after the establishment of the Welfare State education system, mentalities of the class, national, ethnic, and racist types doggedly persist. This fact must be in part attributed to the shortcomings of the system itself, where it has failed to promulgate the openness previously discussed. On the issue of class in particular: where the public system, or at least certain forces in the system, exclusively advocates the values and attitudes of what is usually called the 'working class' kind, it fails its students just as much as does the private system when it exclusively recommends what it calls 'middle' or 'upper class' values and attitudes.

As an unflagging proponent of openness, the state-educated liberal views with concern other deficiencies in the system on which he has relied: the increase in bureaucracy and managerialism, with its impersonal and institutionalised mentality, which is often more concerned with statistics on pages than with the live, complex realities of actual teaching situations. He knows that the bureaucratic and managerial outlook, taken beyond a certain point, is inimical to genuinely liberal education, which has a rich diffuseness that cannot be quantified in any simple, on-paper fashion.

Further, partly in connection with the above tendencies, the liberal is abidingly concerned with intellectual and cultural calibre. His freedom from all forms of collective prejudice makes this inevitable, since his emphasis is always on the individual and the quality of the latter's endeavour and attainment. This clarity of focus—denied to the organisational mind—is one of the great benefits of being *declasse* and non-institutionalised. Also, the clarity carries a sense of excellence, and of gradations below that standard: a sense which the free and unafraid mind should never be without.

It may be objected that a society consisting of people with such strongly developed personal outlooks cannot really be cohesive; but the reply is that, on the contrary, it will cohere more than other societies, since its members will be fully conscious of, and will fully choose, the conditions on which they co-operate. Collectives function better on the basis of clear vision than of blurred; and, in the collective project, the Welfare State liberal can play as full a role as anyone.

Liberty, Equality, Fraternity

The above three words constituted the chief rallying cry of the French Revolution of 1789. This revolution, despite its brutal excesses and its eventual, unintended outcome in the rule of Napoleon, did succeed in inaugurating political modernity in the West: that modernity being, in essence, a refusal to go on accepting autocratic, non-elective forms of government.[1] The words associated with this historic turning-point carry, therefore, enormous weight.

At the same time, they call for close examination. They do not mean the same thing. Liberty is not identical with equality. To be free to do what one wishes to do, either as an individual or as a group, is not to be the same as another individual or group which also has such wishes. The latter differ, from person to person or group to group. Also, to be as free as the next person does not mean having the same abilities as that person. Capacities vary. Further, this difference in capacity means that equality of opportunity to develop one's own potential does not mean equality in the outcome of that development. As has previously been pointed out, equality of opportunity does not lead to post-opportunity equality of condition.

However, there is one sense in which equality should be invariable in all contexts: equality of political, civil, and human rights, and equality before the law. This should be totally unaffected by internal differences between individuals.

---

[1] In this, it is closely linked to the American Revolution against British imperialism in the 1770s.

Moving now to fraternity. The notion of brotherhood is certainly not in conflict with that of liberty; indeed, it is arguable that true brotherhood can only result from the freedom of association, and of expression of feeling, which are integral to liberty. Thus liberty can be seen as a necessary condition for fraternity. But that, of course, does not make it the same as fraternity.

Just as fraternity differs from liberty, so it differs from equality. The concept of the brotherhood of human beings does *not* imply their equality. Just as members of a family are regarded as such even though, as individuals, they differ in aspirations, capacities, and familial rank, so it is, or should be, with members of the human race as a whole. Harmony and co-operation with others do not require sameness. They are perfectly compatible with a sense of variety and even hierarchy.

Liberty, equality, and fraternity remain, after 200 years, key words in political and moral discourse in the West. But they retain this status because their meanings, and relations to each other, are not simple. To think otherwise leads to very damaging consequences, as shown by, for example, the political systems in the 20th century which called themselves Marxist and which claimed descent from the 1789 Revolution. In those systems, there was, at least at an official, doctrinal level, a curious conflation of all three words — producing, or aiming to produce, an homogeneous mass-mentality which abolished awareness of the complexities we have briefly examined.

Anti-Bourgeois Attitudes in the West

[Some points in this essay echo ones made in the essay 'Questions Facing Socialism as a Cultural Outlook'.]

The present (2009) economic crisis in the West, bound up as it is with excessive speculation and risk-taking in the financial sectors of the major Western economies, has aroused a great deal of justified ill-feeling toward the most powerful groups in the world of finance—chiefly bankers. This ill-feeling is yet another reminder of the gap that exists in Western society between the majority of people and the minorities who command enormous economic power: financial, industrial, commercial.

Western society, indeed world society, has of course always been divided between the few who hold major economic power and the many who do not. Prior to the first Industrial Revolution, that power chiefly took the form of the ownership of land and therefore of the means of agricultural production, and was held mainly by a landed aristocracy. With the Industrial Revolution, economic power gradually passed, in the main, to an urban middle class engaged chiefly in industry and manufacture, and in financial activities corollary to these, such as banking, investment, and speculation. This remains the situation today.

These economically dominant groups are collectively described by Marxists, and by a number of other thinkers, as the 'bourgeoisie'. As used by Marxists, the term is pejorative, as one would expect. However, it is also pejorative, generally speaking, when used by those who are not Marxists: a fact which points to considerations of a deeply historical, cultural, and ethical character. While, as said, society has always had economically dominant elites, rarely before has an elite aroused such hostility among cultured people as has the industrial/financial one. This hostility goes back to the early 19th century, and continues to the present day.

Regarding the way it was expressed in the 19th century, let us consider the French writer Flaubert, who was not a Marxist or a socialist of any description, and his novel *A Sentimental Education.* The latter was published in mid-century. It conveys the view that, in the words of literary critic Edmund Wilson,

> our middle class society of manufacturers, businessmen and bankers, of people who live on or deal in investments, so far from being redeemed by its culture, has ended by cheapening and invalidating all the departments of culture, political, scientific, artistic and religious, as well as corrupting and weakening the ordinary human relations: love, friendship, and loyalty to cause — till the whole civilisation seems to dwindle.[a]

Flaubert's view may well seem an exaggerated one, but it certainly finds echoes in the writings of several other 19th century figures who were, like him, not on the political Left. For example, his fellow French novelist Stendhal famously said that he would rather spend two weeks out of every four in prison than in the company of tradesmen. Still another Frenchman, Tocqueville, accused the middle classes of having no conception of intellectual and moral greatness; they were "moderate in all things, except in taste for material well-being".[b] In England, Carlyle protested against the 'cash nexus' basis of relationships in industrial and commercial society. In Germany, Nietzsche averred: "Today, mercantile morality is really nothing but a refinement on piratical morality — buying in the cheapest market and selling in the dearest." He went on to recommend:

> We should take all the branches of transport and trade
> which favour the accumulation of large fortunes, *especially*
> *therefore the money market* [italics mine], out of the hands of
> private persons or private companies, and look upon those
> who own too much… as types fraught with danger to the
> community.[c]

In the 19th century, there were also, of course, Engels, Marx
and Marxists, and other socialists attacking the hegemony of
the bourgeoisie. But Flaubert, Stendhal, Tocqueville, Carlyle,
and Nietzsche are prominent examples of a significant group
of non-Left thinkers who were also engaged in the same
attack. The relevance of their thinking to today's world is
especially clear from the second Nietzsche quotation.
Further, a major strength they all shared was a refusal to
idealise modern industrial and commercial workers, or to
regard this enormous and variegated number of people as,
en masse and in unitary fashion, capable of providing the
solution to the problems presented by the bourgeoisie. This
strength of observational acuteness and critical acumen is
unfortunately not one displayed by many on the Left today.

In the 20th century, anti-bourgeois feeling in the West has
persisted in a variety of forms, though most prominently in
Marxism and on the political Left in general. Outside the
political sphere, it has been voiced by a number of creative
writers and artists of various kinds, especially — again — in
France. One of the most notable of these writers is Sartre,
who, in novels such as *Nausea* and in non-fiction works such
as *Saint Genet,* is clearly out to *epater le bourgeois*. In Britain,
D.H. Lawrence is a stand-out figure, in novels such as *Lady
Chatterley's Lover.* These literary attacks are aimed primarily
not at the economic power enjoyed by the bourgeoisie but at
the psychological and moral smugness widely associated
with the bourgeois mentality.

Some indications have already been given about the roots
of anti-bourgeois attitudes: the view that the bourgeois way
of life is mentally narrow and complacent; that it under-
mines genuine culture and the moral fibre of human

relations; that its god is material well-being. These criticisms claim, essentially, that bourgeois-dom rests on superficiality of mind. This state of mind, it is argued more extensively, knows little or nothing of: ontological and moral anguish, arising from a grappling with first-order philosophical issues; profound and painful psychological struggle; encounter with radical originality in artistic and scientific thought and activity; experience of sustained emotional spontaneity.

Given this picture, and the very negative feelings the picture has produced, we must now turn to the conceptions which thinkers of the last two hundred years have had of pre-bourgeois elites and their respective mentalities.

It is an important cultural fact that in the 19th century there was a distinct growth of interest in mediaevalism on the part of those who were deeply dissatisfied with the prevailing industrial-commercial order. This growth was especially prominent in Britain, in the work of Tennyson, Carlyle, and Morris, and, in the very early part of the century, in that of Scott. Its prominence is perhaps partly due to the fact that, in the 19th century, Britain was the world's most industrialised and commercialised society; hence, there was a sharper sense of nostalgia than in other countries for the pre-industrial past.

Interest has continued in the 20th century, and not just in Britain but in other Western countries, though expressed mainly through the mass-media of cinema and television rather than the written text. (It is possible to interpret the productivity of the mass-media as partly a reverberation of the influence exerted by 19th century texts; this is especially so with Scott's novels.)

The focus on mediaevalism is understandable not only in terms of the nostalgia factor but also because mediaeval culture was the first really distinctive one to have emerged in the West after the Dark Ages, and the one which, extending as it did from the Dark Ages to the Renaissance, has held the longest tenure to date. And the focus is comprehensible in more respects still. Mediaeval culture was regarded as having achieved a truly integrated kind of society, one in which social relations, though rigid, were definite and enduring, involving long-term commitments between individuals and groups: for example, between king and nobles, landowner and tenants, head of religious orders and their members. This view of the stability, or at least relative stability, of such a society compared favourably, in many people's eyes, with what undoubtedly was and is the much more fluid and mutable character of industrial-commercial society, with its largely 'cash-nexus' basis of work relationships, its lack of long-term commitments (especially between employers and employees), and its generally more atomistic nature.

In addition, the culture of knightly chivalry, including physical courage and prowess, was seen as being far preferable to the total absence of such culture in almost all industrial and commercial processes. Now, even allowing for a sizeable element of idealisation in the picture people held of the chivalric ethos, there is no question that the latter was artistically fruitful in its day, inspiring a body of epic and romance poetry which still occupies a major place in Western literature. As well as this poetry there are, for example, the history plays of Shakespeare, dealing retrospectively with a particular part of the mediaeval period in England. The fact that these never flag in dramatic power and vividness is due, not only to Shakespeare's literary genius, but also to the sheer dramatic quality inherent in his subject-matter: the dynastic struggles and related moral issues of that time. One wonders if a modern equivalent to Shakespeare, even with all his predecessor's skills, could ever produce material of comparable interest in dealing with the main figures in British industry and commerce over the last 100-odd years.

Further, it is the case that the industrial-commercial way of life has not inspired any epic literature—indeed, very little literature of any kind which celebrates and endorses it. This fact can be taken to indicate that industry and commerce, while *economically* central to the Western world, are *culturally* peripheral. Thus the West is faced with a deep problem: that sources of economic power are rarely sources of intellectual and emotional elevation, rarely foci of dramatic and moral interest.

Returning to the attention that was previously drawn to the conceptions which various people have had of mediaeval life, and to the kinds of culture which the latter produced: in examining these subjects, there was no wish to evade what, from a modern standpoint, must be regarded as the negative aspects of mediaevalism. Such aspects are numerous, and they include: the rigidity of class divisions; the harshly restrictive economic conditions in which most people lived; the subjugation of women, at all social levels; the absence of democratic political processes; the stunted development of science and technology; the institutional power of the Catholic Church, and, as a related point, the confinement of philosophical thought to parameters compatible with Catholicism. These and other shortcomings must constantly be borne in mind.

They stand in stark contrast to what are—again from a modern viewpoint—the advantages found in modern Western industrial-commercial society. These include: the much larger degree of social and economic mobility; the much improved general standard of living; the increased social and professional status enjoyed by women, and the existence of feminist movements; the operation of democratic political processes; the expansion of human and civil rights; the enormous advances in science and technology; the reduced influence of the institutional powers of all religions, not just Catholicism; the virtually open access to all forms of philosophical thought.

At the same time, it should be noted that I said that these advantages are "found in" Western societies with modern economies, not that they are an automatic product of such economies. Generally, those to do with economic well-being and technological advance are bound up with modern economy. But other kinds of advantage have more complex roots, historically and culturally; for example, democratic politics, the increased status of women, the widening of human rights, the decline of religious authority, the growth of pure science,[1] and of access to the full spectrum of philosophical thought. The existence of benefits of this kind is by no means guaranteed by the mere operation of industry and commerce. It is possible to imagine a state of widespread economic comfort combined with an almost total psychological servility, of the Huxleyan 'brave new world' kind. Also, outside the West, the workings of modern industry and commerce in a country such as China do not carry these benefits with them; while, in other Third World countries, they do not even ensure economic well-being for most people.

Thus it is clear that many of what can be termed the advances of modern society over mediaeval cannot be solely attributed to changes in the economic system; and, to repeat, that system has shown itself to be seriously deficient in contributing to the general culture.

Consideration of its cultural failures returns us to the "deep problem" defined earlier. How can this problem be effectively addressed? Can the continuing anti-bourgeois attitudes be replaced by positions which are constructive and inspiring?

Marxists and other far Left groups will give their own kind of affirmative reply. They will answer 'yes' to both questions by contending that the modern industrial/ commercial working class should — collectively, but through a vanguard political party — seize economic control from the bourgeoisie, and do so outside the sphere of constitutional or electoral process. In gaining control, they will end the

---

[1]   To a considerable extent, this growth can be distinguished from advance in technological efficiency. In Nazi Germany, for instance, pure science in any genuine form was not pursued beyond the level required to underpin industrial and military processes. Indeed, beyond this level it was actually suppressed.

capitalist system and replace it by a socialist one. Industry and commerce will be transformed from systems which primarily seek to make profits for private groups to ones which seek the economic betterment of all members of society. In the process, sources of economic power will become sources of cultural inspiration, and society will attain to a kind of integrated wholeness which is impossible under the dominance of the bourgeoisie. Socialist economics will, in itself and through the changes it produces, lead to the general enrichment of intellectual, emotional, and cultural life: an enrichment which remains out of the question while industry and commerce continue to be run on capitalist lines.

If the above is an accurate precis of the far Left position, then critical comment on it runs as follows:– Given what has already been said about the size and variegated character of that group of people called the industrial/commercial working class, the question can be asked: does this group, as a whole, display a better record than the bourgeoisie in manifesting interest in the topics referred to earlier in the essay, i.e. painful psychological struggle; engagement with ontological and moral complexities and challenges, thus with first-order philosophical issues; extensive concern for artistic originality and creativity? It is vital to ask this question because, according to far Left thinking, the working class *must* lead the way in social and cultural transformation. And if the answer to the question is — no, this group of people does not on the whole display a better record, then serious problems arise for the far Left.

To this point, the far Left may reply as follows: even if it is true that the working class does not perform better than the bourgeoisie in this regard, such has been the case only so far in history, and is so only because the class has not yet had sufficient access to educational and cultural opportunity. However, if this is the counter-argument, the Left still face two problems. One is that they advocate that the working class, as it is *now* and not as it may be in some future social dispensation, should assume control of society: should assume control, then, from an admittedly inferior position educationally and culturally (albeit one that is no fault of their own). Clearly, this does not come across as effective advocacy for such a group's taking the social helm.

Secondly, granted the inadequate cultural performance of the working class, the Left must explain why it is that the bourgeoisie, with all its social privileges and therefore access to cultural opportunity, also generally shows poor performance. In other words, if one large social group (the bourgeoisie) has failed to demonstrate that it *collectively* occupies a position at the cutting edge of human thought, what reason is there to think that a very much larger group (the working class) will ever collectively do so?

These considerations are pivotal because they lead to the issue of deep differences in ability-calibre between *individuals,* of whatever social location (see the two previous essays). This issue points to the difficulties entailed by viewing modern industrial/commercial workers in too generalised and even sentimentalised a fashion. Given such difficulties, the following line of argument merits serious study:– Changes are indeed needed in the ways industry and commerce are run: changes which will bring these activities much more into the general cultural orbit, and produce much wider economic benefits. What is required is the dismantling of all large-scale concentrations of private economic power, and bringing these under an effective form of public ownership and control.

Since the most powerful section of the bourgeoisie can be identified with large-scale formations of economic power, this objective is anti-bourgeois. However, the objective cannot realistically be regarded as attainable by industrial/ commercial workers as a whole, because the vocational capacity and moral dedication required for its success are nowhere near evenly distributed across such a huge number of people. The same is true of the capacity for cultural responsiveness, which is also requisite if the aim is to establish close links between economic activity and cultural life. What would seem to be called for, then, is a group of exceptionally able, committed, and culturally advanced individuals, not to be defined in class terms, who would bring about economic transformation. They would do so through electoral means, and therefore with popular support, of a general kind: again, a support not to be defined in class terms, and consisting of people of all walks of life and backgrounds, people who wished to end the economic dominance of one group over others.

If this argument is accepted, then there is manifestly a need, in our ways of thinking positively about future social directions, to transcend attitudes which are underpinned by class concepts—certainly rigid concepts, but even, in the long term, all such concepts, at least as traditionally used.

The termination of economic dominance, and its replacement by a system in which private interests played only a minor economic role, would hopefully produce a situation in which people saw each other in individual rather than group terms. Indeed, the maximising of people's clarity of vision, in regarding each other as individuals rather than as members of social sectors, might be described as the ultimate destination of all social progress and reform, throughout the whole of human history. With this maximisation, there would be no 'anti-bourgeois attitudes', or anti-class attitudes of any kinds. Oppositional outlooks, where they existed, would be only between individuals, or groupings of individuals who never allowed their personal identities to be submerged under any kind of collective definition. Such has in fact been the case at all higher cultural levels throughout history, especially in philosophy and the arts, and there is no valid moral reason why it should not become the case right across the modern social spectrum.

[a]   Edmund Wilson, *The Triple Thinkers*: London, Pelican Books Ltd., 1962 (1952), p. 95.
[b]   See Alan Kahan, Aristocratic Liberalism: The Social and Political

Thought of Jacob Burckhardt, John Stuart Mill and Alexis de
Tocqueville: Oxford, Oxford University Press, 1992, p. 49.
[c]   For both quotations, see Will Durant, *The Story of Philosophy*:
London, Ernest Benn Ltd., 1955 (1947), p. 375.

French Rationality in the 18th Century

It is arguable that France was, on balance, Europe's intellectual leader in the 18th century. Though France has had other periods of philosophical eminence, notably with Descartes in the 17th century, Comte in the 19th, and Sartre in the 20th, there is a case for saying that in no century apart from the 18th did it possess such a large number of important thinkers, not all front-rank certainly, but all, in their combined effect, creating an overall impact greater than that of any other national group in Europe at the time. Voltaire, Diderot, D'Alembert, D'Holbach, Condillac, Montesquieu, Helvetius, Condorcet, and Turgot: no other European country in the 18th century can match this cluster,[1] even if certain other nations produced individual philosophers who were arguably of greater stature than any in this group, e.g. Berkeley in Ireland, Hume in Scotland, and Kant in Germany.

All in the group were noted for their rational acuity. Criteria implied by this phrase include the following: logical coherence in argumentation; clarity in moving from premise through to conclusion; and, where possible and appropriate, reference to available scientific data in support of all stages of reasoning. All the thinkers in question displayed effective approaches to a wide range of philosophical issues. Bearing in mind the period in which they were writing, it should be noted that only some of these issues were bound up with the problems of living under the repressive *ancien regime* which

---

[1]   It will be noted that the list excludes Rousseau. Rousseau was undeniably a major 18th century figure, but our focus is on *rational* thought, and Rousseau cannot be regarded as primarily a spokesman for rationality.

held sway in France until 1789. In addition to topical political concerns, there were many others, linked to perennial themes in philosophy. Such eclecticism marked the French thinking of this period as, in fact, a major component of the general Enlightenment process, which gathered strong momentum in the 17th, stronger in the 18th, and went on to reach even greater heights in the 19th and 20th.

The above reference to the 17th century needs extending, because of the intellectual relevance of that period to the century which followed. Descartes has already been mentioned as a representative of French thought; Leibniz and Spinoza can also be referred to, as two other leading Continental figures. However, when it comes to looking at one country which made the largest national contribution to European thought during this period, that country is surely England. Bacon, Hobbes, and Locke furnished a massive contribution to the Enlightenment by carrying the empiricist school of thinking into radically new areas. Empiricism is the view that knowledge of the world is based on experience of the world. This pioneering effort was linked, not surprisingly, with the growth of science in England; Newton and Boyle were among the luminaries in the scientific field.

Also in England during this century there were major political developments. The challenge to unbridled monarchical power, which had been gathering force over a long period of time, came finally to a head with the Civil War of 1645–9, fought by forces supporting parliamentary institutions and those supporting the crown; the former being the eventual victors.

Philosophical and political happenings in England were to have a deep impact on the French mind in the 18th century. The independent, questioning spirit which fuels empiricism was to give new vigour to French thinking, and the military victory over monarchical hegemony was to set an example of how such power could be overthrown. Admiration for general English achievements was extensively voiced by, for example, Voltaire; and, among English philosophers, Locke (as we shall see) was especially influential on French thought.[2]

Let us begin with Voltaire (1694–1778), France's leading cultural personality during this period. Voltaire was a polymath—playwright, novelist, and historian as well as philosopher. Deeply influenced by Locke's empiricism and by England's liberal political institutions, he expressed his Anglophilia in *Letters Concerning the English Nation* (1733),[3] a work which contributed to the development of liberal thought not only in France but also on the Continent as a whole.

Voltaire's most extensive philosophical work is the *Philosophical Dictionary* (1764), in which he expressed his views on metaphysics, religion, politics, and ethics. His other main achievement in philosophy is actually a work of fiction, *Candide* (1759), which attacked the cosmic optimism of Leibniz, as expressed in Leibniz's dictum that the world is "the best of all possible worlds". (Though Leibniz's optimism emanated from religious belief, Voltaire's criticism of it does not mean that he himself was a non-believer. He was in fact a Deist, but not, like Leibniz, a Christian.)

Voltaire also contributed to the *Encyclopaedia,* an enormous work compiled from 1751–65 under the directorship of Diderot and D'Alembert.

Diderot (1713–84) was, like Voltaire, a man of many talents, and, again like Voltaire, influenced by Locke. To the *Encyclopaedia,* he contributed articles on aesthetics, ethics, social theory, and the philosophy of history. Specifically in philosophy, he was a materialist and, echoing Locke, examined the influence of sense experience on the acquisition of ideas. He regarded observation and reflection as playing complementary roles in empirical research. He saw experimental science as being possible only because a single

---

[2]   Locke was also, of course, influential on American thinkers. His ideas were a shaping factor in the drafting of the American Constitution in the 1770s.

[3]   This was later expanded to *Philosophical Letters* (1734)

causal principle was at work in nature. Material substance was, he argued, composed of molecules. Also, he held an early form of evolutionary theory: all species of living things went through stages of development. Finally, in the field of psychology, he postulated that the formation of values was derived from childhood influences (this anticipates Freud). His main philosophical works are *The Dream of D'Alembert* (1730) and *Thoughts on the Interpretation of Nature* (1754). Manifestly, Diderot is a major figure. His perspectives on empiricism, materialism, biology, and psychology remain highly cogent ones.

Diderot's co-director on the *Encyclopaedia*, D'Alembert (1717–83), did his most distinctive work in mathematics, giving his name to a vital principle in theoretical mechanics. At the same time, he was a man of wide philosophical and literary interests.

Sharing affinities with Diderot was D'Holbach (1723–89). In his main work, *The System of Nature* (1770), D'Holbach argued for an absolute naturalism. There was a fundamental continuity between man and the rest of nature, between animal and human behaviour. Also, the natural world was knowable through human experience and thought, rather than through traditional religious beliefs and alleged 'revelations'. All natural phenomena, including mental ones, were explicable in terms of the organisation and activity of matter.

Combined with this naturalism was an ethical eudaemonism and political liberalism. The chief moral aims of man should be happiness and self-preservation. To attain both, accurate knowledge of nature was indispensable. Further, nature made man neither good nor bad but adaptable through education and experience. Reason indicated men's need for co-operation with each other, and was the basis of moral systems, whose justification was social utility. Politics should conform to general social aims, and not to those of powerful individuals or elites. The power of man over man was defensible only in terms of general social utility. Finally, education and legislation could be effective only when people were convinced that these activities genuinely served their interests.

Echoing Diderot's, this perspective affirms materialism, upholds empiricism, and highlights biological considerations. It also explores how behaviour is influenced by external factors. Moreover, it recalls Voltaire in its advocacy of liberal practice in social and political affairs. D'Holbach's continuing importance in philosophy cannot be in doubt.

The materialism of D'Holbach and Diderot found a further voice in Lamettrie, who added to the doctrine the view that matter was not inert but perpetually in motion (compare the later doctrine of dialectical materialism). On this view, there was no need for a divine prime mover, contrary to what deism and other theological positions argued. Also, mentality was dependent on, and a function of, the material world.

The determination of mind by matter — again, D'Holbach's view also — is a postulate that was to be very influential in later philosophy. From it, Marx, in the 19th century, derived the theory that mind is a by-product of bodily organisation. Further, it is the basis of modern epiphenomenalism, which shares Marx's position but adds that mind is always a passive by-product; see in particular the philosophy of Santayana in the 20th century.

Leaving ontological considerations to do with matter and mind: the philosophy of Montesquieu (1689–1755) was entirely political and juridical. Another thinker influenced by Locke, Montesquieu was concerned to define different kinds of governance, and to achieve this he looked for what he called the "animating principle" in each kind. As regards despotism, which was what many 18th century Frenchmen saw the *ancien regime* as being, the animating principle was fear. In monarchy, at least the kind that was not despotic, the principle was honour. In republican and democratic systems, it was virtue. From a modern standpoint, Montesquieu's ideas about democracy are clearly the most salient ones; and, in this area, his Lockean focus on the idea of separation of powers between the legislative, executive, and judicial arms of government is obviously germane to contemporary democratic systems, especially that of the United States.

Yet another adherent of Locke was Condillac (1715–80); though with him, as with Diderot, the emphasis was epistemological rather than political. He agreed with Locke in tracing all human faculties back to their origins in sensations. Man's will and understanding were modifications of the accumulations of impressions and associations arising from the stimulation of the sense-organs.

Locke's impact was felt still further in the work of Helvetius (1715–71). Helvetius emphasised the function of experience in our acquisition of knowledge, and regarded sensation as the basis of cognition. Also, like Diderot, D'Holbach and Lamettrie, he was a materialist.

There are several other interesting aspects of Helvetius's thought. He echoed Hume in his own century, and anticipated Schopenhauer and Nietzsche in the 19th, in contending that specific interests and passions were the mainspring of our striving for knowledge. Further, he sharply criticised the Catholic Church for inculcating what he regarded as erroneous moral ideas. Also, he called for a reform of the educational system in the interests of individual and social emancipation, and encouraged everyone to participate in the search for truth and enlightenment. He equated truth with the moral good. Finally, in stark contrast to Rousseau, he welcomed further developments in all aspects of culture and civilisation: science, industry, art. Overall, Helvetius's outlook was an extremely well-rounded and progressive one, highly pertinent to us today.

Lastly, let us briefly consider Condorcet and Turgot. Condorcet (1743–94) was a man of wide cultural and historical perspective. He pioneered the mathematical and statistical study of social institutions and trends, especially in connection with electoral systems. His work is extensively relevant to modern democratic politics. Turgot (1727–1781) was primarily an economist. He was an early advocate of economic liberalism and called for the complete freedom of commerce and industry. Including but also going beyond economics was the all-inclusive view he held of human progress, a view which covered not only unhindered economic practice but the whole of culture: the arts and sciences, social mores and institutions, and legal codes.

This survey of rationality in 18th century France is, though brief, an indicator of the kind of contribution that country was making to Western thought two centuries ago. As said in the introduction, the collective achievement was arguably without equal in its day. That is partly because it was emphatically not confined to a consideration of topical political issues and problems, extensive though those were. We could go on to say that the national attainment compares very favourably with those of England in the 17th century and of Germany and England in the 19th.

Also, we should note that it emerged almost entirely while the *ancien regime* was still in place. This fact denotes that great cultural achievements are clearly possible under conditions of political repression—indeed, may arise partly *because* of that repression. A similar phenomenon, though literary rather than philosophical, is to be found in 19th century Czarist Russia.

A case can be made for saying that France in the 19th and 20th centuries failed to reach the philosophical heights it had scaled in the 18th, but then that leaves the question: will it do so, or do even more, in the 21st?

Mental Perspectives
in the West

Several thinkers, including (famously) Spengler, have claimed that Western culture has long ceased to be a great originator in the world, and has lapsed into a preoccupation with material comfort and security. Hence, they argue, it has ceased to give the world new mental and moral horizons. This view, though clearly an exaggeration, deserves attention.

If we date modern Western culture approximately from the Renaissance, then we certainly see what past achievements have been in extending general human horizons. For example, the 17th century witnessed the emergence of modern scientific method, built on a legacy from ancient Greece which had been largely neglected in the West for the previous thousand years. On this method, all subsequent scientific activity, world-wide and whatever the cultural context, has been based.

Partly in connection with this scientific efflorescence, there was the growth, from the 17th century onward, of new forms of philosophical thinking which, unlike the mediaeval philosophy that had preceded them, were not dependent on authority of any kind,[1] either Christian or classical. Whether new developments in empiricism or rationalism,[2] these

---

[1]   Thus Descartes, with his overt rejection of all previous authority, is rightly regarded as the 'father' of modern Western philosophy.

[2]   On empiricism, see the preceding essay.
       On rationalism: the view that knowledge about the world is derived from purely logical and rational thinking prior to experience of the world. This thinking, according to rationalism, shows how reality logically has to be constituted.
           (Given this definition of rationalism, a distinction must obviously be drawn between it and the term 'rationality', as used in the previous

novel directions displayed an unprecedented autonomy,[3] so setting a future example for the rest of mankind.

A third development, and closely linked to the preceding, was a radical questioning of absolutist forms of government, and the formulation of arguments for popular representation and democracy. In this, as has been said previously, England led the way. It was to be followed by the United States and France in the 18th century, and subsequently by many other countries across the world. The overall global impact of the democratic principle is a point which hardly requires making. Now, as never before in world history, all progressively minded people criticise a government for not being democratic; never before has failure to meet democratic standards been a cause of such widespread opprobrium.

These, then, are among the West's intellectual and moral achievements, and bequests to the rest of mankind. They have a truly heroic dimension in that they represent victory over opposition — opposition that was especially fierce when directed against efforts to attain scientific and political progress. Huge risks were taken, immense courage was displayed, and, chiefly in the political context, many people perished for the cause. Ideas, principles, and visions were a basis for the sacrifice of security and of life itself.

Before viewing these in relation to the contemporary West, it is necessary to set them against the main short-

---

essay. At the start of the second paragraph of that essay, the term 'rational acuity' was deployed, then expanded on. This way of defining rationality as a wide-ranging approach to describing the world clearly differentiates the term from 'rationalism'.)

    For both schools of thought, the 17th century was decisive in establishing their modern contexts. Regarding empiricism, Bacon, Hobbes and Locke were seminal figures, as has been said previously. With rationalism, the key figures were Descartes, Spinoza and Leibniz.

[3]   This independence did not necessarily mean that Western thought severed attachment from established doctrines such as religions. For instance, Descartes, Spinoza, and Leibniz all held religious beliefs. But it did mean that religious beliefs were now seen as vindicated only by rational thinking of a completely self-standing and autonomous character.

coming displayed by the Western world, from (ironically) the 17th century to, in fact, the present. This shortcoming has been imperialism: initially in a mercantile form and then in an industrial one. The beginnings of European empires date from the late Renaissance, and the related practice of slavery from soon after this. Portugal, Spain, then Britain and France, and, to a lesser extent, Holland, Germany, and Italy: these have been the main European countries involved.

It is true that, for some Europeans, there was a genuinely held view that extension of Western power into the non-Western world was a civilising process, and that the task of civilising was one that Europeans ought to assume. This feeling was particularly strong in some circles in Britain and France. However, while acknowledging the sincerity of this position, we must also recognise that it was not held by many other colonisers, whose motives were the simple ones of material acquisitiveness and power. Indeed, most people on the political Left see Western imperialism as having been chiefly driven by the sectional interests of the economically dominant groups in the various countries concerned. Obviously, there is much substance in this view, even though it can be said to underrate the genuinely ideological factors referred to above.

From a general historical standpoint, what is curious about the growth of European imperialism from the 17th to the 20th centuries is that it coincided with the positive developments we have previously examined. This implies that, in the countries concerned, there were many *different* forces at work; some in stark opposition to others; some influential in one area of society, others in another. Adequate commentary on the situation could come only from a highly detailed historical study, one which is beyond the scope of this short essay. For now, suffice it to suggest that, in addition to the points already made about varying attitudes to imperialism, it was the case that, for a sizeable number of people, the progressive thinking we have looked at remained *Western-centric*. For these people, it applied only or mainly to the West, with conditions in other parts of the world being seen as largely outside the sphere of relevance. In other words, the progress of science, independent philosophical thinking, and political emancipation were viewed as chiefly an internal Western affair.[4]

As regards the present day, many people, especially on the Left, would regard American foreign policy as the most prominent current form of Western imperialism—having gradually superseded, post-WW2, that of the West European powers previously referred to. Also, U.S. policy is widely seen as having been, on balance, more aggressive than that of the Soviet Union during this period. Thus American incursion into Indo-China in the 1960s and 70s, and currently into Iraq and Afghanistan, is interpreted by the Left chiefly as a drive to control natural resources and to secure a dominant geo-strategic position in these areas. Again, the influence of economically dominant groups is taken, by the Left, to be the main factor in shaping this policy.

The Left's argument is certainly a strong one. To those who retort that the principal factors behind these policies were and are primarily *not* economic and strategic but moral and ideological, the following needs to be pointed out: firstly, a genuinely held anti-Communist doctrine can certainly be seen as one of the forces actuating policy in Indo-China, but whether it was the main one is open to question. Secondly, if we assume that ideology is bound up with policy, then we have to note that this ideology has radically changed. Anti-Communism can no longer be used as an argument in the case of Iraq and Afghanistan. Is the

---

[4]　It might be added that this Western-centrism, as a cultural tradition, is strongly reflected in a comment made by even such a progressive and eclectic thinker as Russell, and as relatively recently as 1946. Russell argues, in a radical tone, that after World War Two, the West "shall have to admit Asia to equality in our thoughts" (*History of Western Philosophy*: London, Unwin University Books, 1975 (1946), p. 395). In saying this, incidentally, he seems to be unaware that one of the greatest Western philosophers of the 19th century, Schopenhauer, had already, and with emphasis, admitted Asian thought to such equality.

new ideology, then, that of furthering democracy and liberal values, wherever in the world they are needed? This is what was, among other things, invoked as a justification for the invasion of Iraq, but the claim has to be set against the fact that the U.S. staunchly supported Iraq's dictator, Saddam Hussein, for virtually the whole of his political career until 1991, when for the first time he disobeyed Washington. Regarding the current state of 'democracy' in Iraq under U.S. occupation, not much of a positive nature can be said.

In connection with Afghanistan, the argument about democratising and liberalising Afghan society is in conflict with the fact that, in the 1980s and 90s, the U.S. gave backing to groups in Afghanistan who were, at least by Western standards, totally non-democratic and illiberal. These included the Mujaheddin and the Taliban—the latter being, of course, the people whom the U.S. is now fighting. While it is true that this backing was a strategem against Soviet occupation of the country, still it was support, and remains an ineffaceable historical fact which casts a huge shadow over American arguments about their wish to inculcate democratic and liberal values. Such labyrinthine practice suggests different intentions. As with Iraq, when it comes to considering the current state of 'democracy' in Afghanistan, enthusiastic comment is difficult.

The above points about U.S. foreign policy are extremely cogent ones. If accepted, they reinforce the contention that imperialism, the West's main shortcoming since the Renaissance, remains unfortunately operative.

Returning now to the positive aspects of Western culture and history, the question arises: how does the contemporary West measure up to the achievements of its past?

In terms of scientific development, attainment has of course been ongoing. In applied science, the present dwarfs the past, and has transformed our whole way of life, on a scale without parallel in history. This has been especially true in the fields of medicine and general technology. Cures for a host of previously incurable diseases; organ transplants; general health monitoring systems; supersonic air travel; nuclear power stations; space exploration, manned and unmanned: these are among the conquests of nature marking the modern era in the West, especially since 1945.

In these fields, the rest of the world has sought to follow. However, the ability to do so has rested largely on economic factors, and here we come to a crucial consideration. The West leads the world economically, and by a margin which has increased, hugely, in the last 60 years. Its dominant position, and especially that of the U.S., has enabled it to resource scientific projects to an extent which nowhere else in the world has been capable of matching.

This dominance is of course long-standing. In the past, it was inextricably bound up with territorial imperialism and control of material resources, on the part of European powers. In the case of the U.S., imperialism, as has been previously argued, remains a major factor, though the type is only partly territorial. Imperialism aside, dominance in its current form is based largely on unfair, one-sided trade agreements with Third World countries. See in particular the trade deals established by the European Union with other parts of the world.

The economic consideration is linked to vital political issues. We should recall the view expressed in the opening paragraph of this essay: that the West has lapsed into a preoccupation with material comfort and security. Certainly it is the case that most people in the West wish to hold on to the economic superiority they now enjoy (without, in general, enquiring in detail into exactly *why* they possess it). They wish to maintain their standard of living, which, even for the less well off, is high compared with standards in most other parts of the world. This means that, while Westerners accept democracy in its political form, most, from a global standpoint, do not accept it in its economic form. Economic democracy can be defined as serving the will and interests of the majority in economic terms (just as political democracy is the same serving in terms of the kind of government the majority seeks). On a global scale, the majority consists mainly of course of non-Westerners; hence economic democracy on a global scale would do away with the economically exclusive and privileged positions which Westerners now occupy and which, to repeat, most wish to preserve.

It should be added that most in the West do not even accept economic democracy on a *local* scale—that is, within their own societies. They acquiesce in the power of economically dominant elites, which retain their position no matter which party is elected to government.

Acceptance of economic democracy, globally and locally, would of course involve a radical belt-tightening on the part of the hitherto very wealthy minorities in Western society. But this should be seen as a price worth paying for the widening of economic equity and justice. Finally on this point, equity does not mean complete levelling. All that is asked for is the doing away of large differences in economic status, and the effort to achieve economic similarity, not sameness.

Returning to political democracy: although Westerners do accept this, they frequently show less interest in participating in it than one might wish for. Turn-outs in elections in many Western countries have, in recent decades, been often disappointing. In a truly vibrant democracy, the turn-out should regularly be in the 80–90% range; it has rarely been so in the recent past. When one reflects how, in the more distant past, especially in the 18th and 19th centuries, so many people risked life and limb to secure the vote, then the difference with the present is both clear and disappointing.

It would seem that, in the contemporary world, the elementary battle for democracy, whose theatre was once the West, is now being fought mainly outside the West, in countries such as Iran, Zimbabwe, and Thailand. It is of course true that, in countries such as these, the experience of democratic process is relatively new, and the dangers threatening its preservation are much larger than they are in the West. Probably for these reasons, democracy is much more of a pivotal issue than in the Western world: hence the "elementary battle" for it. However, this battle contains a drama and poignancy which it also once had in the West. For that reason, it should be followed with keen interest by Westerners, and with active support for genuinely democratic forces in these countries. Westerners should wish democracy *anywhere* in the world to be entrenched with the relative security with which it is entrenched in their own societies. In other words, there is a need to be democratically inclusive about democracy, the proper functioning of which can be regarded as perhaps mankind's greatest political achievement. Further, as regards their own democracies, Westerners should never be complacent, or cultivate a false sense of security. Wherever democracy exists, it is always a pivotal issue. As the traditional saying goes, 'The price of liberty is eternal vigilance'. Given the influence exerted in Western countries by economically dominant groups, there is every reason to remain eternally vigilant. The danger to democracy may not be as great as in non-Western societies, but it is indubitably present.

There are of course groups in Western society who are fully and actively in favour of maximising political democracy around the world, and of establishing global and local economic democracy. Combining both projects, they are the true progressives; and, in the sphere of organised politics, are by no means confined to one party or political formation. Numerically, they are a minority; yet it is to them that one must primarily look for exemplars of the moral passion that went into Western politics during its great periods in the past. They are the ones who are, for example, highly eloquent and illuminating on the subject of the complacency, fickleness, and narrowness produced by the acquisitive and consumerist mentality which is so widespread in the West. They are the ones who can best show their fellow citizens how to link hands with the citizens of countries beyond the West, in amity and mutual endeavour.

Generally, the majority in Western society needs to be more mindful of, and sensitive to, the economic plight of most of the world's population. Activist minorities such as the ones we have referred to can certainly help them to gain this awareness, but, once gained, it should be a permanent part of their mental furniture. This will enable the majority to see things in clearer perspective and proportion, making them far less susceptible to those forces in society which seek to promulgate the consumerist mentality. Only then will such things as the ownership of a second car, the taking of a second expensive holiday each year, and the buying of the most fashionable clothes or gadgets, cease to be important objectives.

It must be remembered that the great political and moral ideas which emerged in Europe in the 17th and 18th centuries, and which were to affect the whole world, did not arise from generally affluent conditions, just as so many of their predecessors in ancient Greece did not. Perhaps, for a new moral flowering in the West, a leaner, less affluent and comfortable economic situation, across all sections of society, may be required. Such a situation would hopefully create a larger sense of global kinship with the rest of humanity.

Finally, leaving economic and political issues, and turning again to philosophy, we need to consider, in very general terms, the current state of Western philosophical thought in relation to its past. Manifestly, this is a hugely complex subject. Even so, two broad generalisations can be legitimately ventured.

The first is that, in connection with science and technology, the empiricist school of thinking has been greatly strengthened, at the expense of the rationalist. Increasingly, the project of describing reality has been wedded to science, chiefly physics, and scientific philosophy has become the order of the day. Indeed, philosophy has largely given up any claims to possessing methods of describing reality other than those of science; thus, in the descriptive project, it confines itself to critical consideration of scientific methods and modes of thinking (i.e. philosophy of science) and to the formulation of hypotheses which can then be subject to scientific analysis. All this is a far cry from the days when the word 'philosophy' held pride of place in connection with scientific activity—when, in fact, the latter was termed 'natural philosophy', and this well into the 18th century.

Secondly, as empiricism has developed it has increasingly come to view reality in a piecemeal and pluralistic fashion, abandoning attempts to construct all-embracing systems purporting to cover everything in the universe. This is very different from the philosophical endeavours of earlier periods, especially those of the 17th century, which was very much an age of system-building. Modern empiricism is content to proceed as science proceeds, constructing description bit by bit, tentatively, and always aware that much that is unknown lies outside the picture it has thus far fashioned. This approach would, if globally adopted, serve as an effective antidote to the re-resurgence of religious fundamentalisms and fanaticisms which we have unfortunately seen in recent decades.

Lastly, though this approach marks a major departure from the thinking of previous periods since the 17th century, it nevertheless stands on the shoulders of those earlier vigorous and independent endeavours. Without the latter, which broke decisively from mediaevalism, it would not have materialised at all.

Genius & Changes in Social Context

Genius can be defined as the display, within any given discipline, of the widest perceptual/rational/emotional magnitude. This display includes the possession of outstanding communicative skills. Genius, then, is the evincing of exceptional range of vision, and exceptional technique for conveying that vision.

All the epithets used here clearly imply that genius is extremely rare. This implication is surely borne out by an examination of the histories of the various disciplines in the humanities and sciences. Even allowing for the probability that, at various points in the past, great ability went unrecognised, the instances of recognised greatness are sufficiently infrequent to suggest that those of unrecognised greatness were infrequent also. Even if, for the sake of argument, we were to say that, for every great figure who has been recorded, there are two others unrecorded, and that, therefore, the textual coverage of, say, Western philosophy should really be three times as long as it actually is, such an amended text would still be a relatively brief amount of material. While covering two and a half thousand years of Western thought, it would still come to far less than the page total of less than six months' production of tabloid newspapers in Britain; and it would deal with a very small minority of the West's population.

Considerations such as these inevitably lead to the metaphor which has been traditionally deployed to differentiate the genius-level of attainment from other levels: that of the mountain range in contrast to the plateau or plain. The metaphor is effective because it conveys the ideas of height and of rugged longevity: in other words, of enduring influence, constant resonance, fresh appeal to new generations—all of which are bound up with the perceptual/rational/emotional scope previously referred to.

These points now need to be considered in the light of the fact that social contexts change. What is enduring and ever-relevant is juxtaposed with what alters and varies. Society undergoes political upheavals, alterations in economic system and social structure, fluctuation and variation in artistic convention and fashion. Another change, as important as anything else, is the growth in scientific knowledge. What, then, are the achievements of genius in relation to these variations? In attempting to answer this question, we will limit ourselves to the Western context.

Let's begin with looking at the expansion of scientific knowledge, and focusing on the hard sciences: chiefly physics and chemistry. For those who regard science as developmental in a clear linear fashion,[1] a process of steadily accumulating knowledge, then the attainments of genius are milestones along the path of discovery and clarification. Among those milestones are Aristotle, Galileo, Newton, Darwin, and Einstein. These and others have reached levels of perception and rationality which have been the summation of previous efforts combined with absolute originality, and which have opened the way to new levels to be reached in the future. These men have both received a heritage and created one. Thus scientific genius is at one with change, because it is at one with the progress of knowledge. The genuine scientific spirit is completely open and adaptable.

In philosophy, the situation regarding change is not quite the same. While Western philosophy has, as previously said, increasingly referenced itself to advances in scientific knowledge, it has remained concerned with traditional philosophical issues, some of which pre-date the advent of

---

[1]   As distinct from the view that science does not move forward in factual discovery but only changes the mental models or paradigms which it adopts in its attempts to interpret reality.

modern science in the West since the 17th century. Even in the fields most closely associated with science, traditional debates have continued. These include: the nature of mind and matter; the question of whether matter is more fundamental than mind, or the reverse; the question of whether causation actually exists, and, if it does, whether it is mechanistic (non-purposive) or teleological (purposive),[2] and whether it is totally pervasive; finally, the question of whether the human mind can truly know the world external to it, and, if it can, to what extent it can.

In modern scientific philosophy, philosophy, to repeat, claims no method separate from science for investigating reality. But where it is distinctive is in concerning itself mainly with the most general scientific findings, regarding the broadest features of reality discovered thus far. This is what the issues listed in the above paragraph are about. Philosophy's study of the most general aspects of things, rather than with particular, small-scale findings, is ontology. Thus, in ontology, particular discoveries are noted but then fed into the larger, general areas of consideration. And, once again, many of these areas are of long standing.

Further, in addition to questions concerned with ontology, there are other ones concerned with its sister-subject, epistemology: the study of the process of acquiring knowledge. There are other ones still, connected with ethics and aesthetics (the latter being the study of the standards and logic of our judgments of taste and beauty in the arts). Ethics and aesthetics are little reliant on the development of scientific knowledge. Ethics pays less attention to scientific discoveries and more—much more—to the question of the relation between fact (the field of science) and moral value. One of its chief concerns is what to do with what we know, whatever it is that we do know: how to act in the light of knowledge, and how to use knowledge. Aesthetics too has

---

2   More will be said on mechanistic and teleological causality in a later essay with this title.

its own sphere of questions, separate from the possession of scientific knowledge.

In whichever general mode or modes the philosophical genius works, s/he achieves an exceptional degree of precision and coherence within the area(s) concerned. Hence, for example, Plato, Kant, Hume, and Schopenhauer staked out key positions in epistemology and ethics; Aristotle, Leibniz, and Descartes in ontology; Aristotle (again), Spinoza, and Spencer in ethics; and Kant, Schopenhauer, and Croce in aesthetics.[3] In doing so, these figures created exemplars for future thinking in the modes concerned, with new discoveries in science being, where appropriate, referred to these modes.

Moving from philosophy to the practice of the arts, the first issue to be considered is as follows: in what senses are the contents of the arts—literary and visual—perceptual and cognitive?[4] Leaving aside Plato's controversial answer that they are not, in an ultimate sense, perceptual at all, we can reply that they do describe reality, but *not* in the ways science describes it. Science delineates things in the world at a level which lies beneath that of the mental images and sense-impressions of the perceiver: it delineates basically in physico-chemical terms. But the arts are content to remain on the imagistic and sensuous level; this level is their chosen mode of perception and description.

From these considerations, further ones arise. Science's task is to discover laws which will ultimately embrace all the

---

3   Obviously, this list is very selective. Many more names could be given as exemplars in each of the modes, but space restrictions prevent this.

4   It will be noted that music has not been referred to. This is because music is not a cognitive art, in the sense that it does not describe or represent physical objects in the external world. It does express inner emotional response to these objects, but that is a different matter. Also, while it can reproduce acoustic experience of the physical world, the 'sound' on which this experience is based is not actually an external physical object, but an inner, subjective sensation felt in response to the impact of air-vibrations. Air-vibrations themselves can be described as external physical objects, but this description cannot apply to the sensations they produce.

event-regularities found at the sub-imagistic and sub-sensuous level of apprehending reality. Indeed, a number of scientists and philosophers contend that this level is in fact the only one that is, potentially at least, exhaustively descriptive of reality. Moreover, they say that this point applies not only to objects of perception but also to the processes by which these objects are perceived. Hence, mental images, as well as all mental processes and all sense-experiences, are actually physical and chemical states and processes in the brain. As such, they can, just as much as the objects of perception, be fully re-described in terms of event-regularities at the physico-chemical level: the level to which scientific laws apply. The position being articulated here is physicalism.

In accordance with physicalism, it can be posited that a kind of manifestation of these physico-chemical regularities is to be found in certain types of art, chiefly literary but also visual. That is: art which depicts certain enduring features of human experience thus far, certain ways of (to use non-physicalistic language) thinking, feeling, and interacting with others which have been recurrent throughout human history, or at least for very long periods of time within that history. These recurrences are themselves a form of regularity; and, on the view that they can be more fundamentally described in physico-chemical terms, they are constituted by regularities of the latter type.

The delineation of these recurring forms of experience[5] has reached a definitive and archetypal level in the works of the West's greatest creative writers, who include Sophocles, Dante, Shakespeare, Dickens, and Tolstoy. It is found also among the West's greatest painters, who include Leonardo, Michelangelo, Raphael, and Rembrandt. The level is archetypal and of enduring value because, as art, it depicts thoughts, feelings, and things in the world solely at the level of mental images and sense impressions, and therefore

---

[5]   And recurring, we must assume, for as long as humanity remains in its present state of physical evolution.

cannot proceed to other modes of description. These other modes are of course the scientific ones which we have been discussing, and which are, for physicalists, the more fundamental. Hence, in *not* going beyond the sphere of the mental and the sensuous, artistic depiction achieves a kind of final status in its own terms, if the depiction cannot be aesthetically bettered. In this way, it attains the stature of a 'classic'. This is in obvious contrast to the sciences of physics and chemistry, whose descriptive content changes when new discoveries are made. These changes continually add to our understanding of the experiences and the things in the world which classic art can portray in its own kind of 'final' way. In science, then, there is no such thing as a once-and-for-all, classic descriptive moment.

Now, in broad relation to what has been said about science, philosophy, and art, we must return to a consideration of changing social contexts.

In science, as said, change is of the essence: it is intrinsic to the scientific project. But of course science can produce change not only within its own sphere but also beyond it; its discoveries and inventions alter society as a whole. At the level of invention and technology, this has emphatically been the case in the West since the first Industrial Revolution. To the great inventors, such as Watt, Stephenson, Faraday, Diesel, Edison, Baird, and Marconi, fundamental changes in way of life can be credited.

Philosophy too has contributed to major social change. The growing linkage between philosophy and science since the 17th century, plus philosophy's increasing engagement with progressive political and social developments during the same period, have been key factors in the West's transition from mediaevalism to modernity. The latter term — a very broad one — includes reference to the following points, all of which have been made earlier in this book: the growth of an empirical, science-oriented way of thinking, one independent of political or religious authority; the consequent questioning of religious doctrines, in relation not only to ecclesiastical organisations but also political and social systems; and the democratic impulse toward inclusion of more and more social groups in the apportioning of political and economic benefits. Partly as a result of philosophy's input, one to which its greatest figures have made crucial contributions, the Western world of today is an incomparably more open one, intellectually and politically, than it was before the 17th century.

As regards the arts: it has already been said that the literary and visual arts are not developmental in two, related, senses: 1) works of genius achieve classic status, which sets definitive standards for posterity; 2) these works attain such status partly because they remain at the imagistic and sensuous levels of description.

There is also a third sense in which development does not occur, and this includes music and all the other arts not previously examined. Styles, conventions, and fashions have varied greatly across the centuries, but this variation is not itself a form of progress. The transition from one style to another is not a kind of linear advance; it is only change. It is true that the style of one age may inspire and influence that of a later period, but again this is not a matter of progress.

Given this amount of formal variation, the following questions arise: what is the relation between, on the one hand, style and convention, and, on the other, content? If artistic productions are viewed hierarchically, as they indeed are when given classic or 'masterpiece' status, does it follow that hierarchical judgments can only be made about content, if forms and conventions in themselves are neither superior nor inferior to each other?

The answers to both these questions can be summed up in the statement: greatness is a matter of content-cum-form. In other words, the perceptual/rational/emotional scope of a work is the primary consideration, and of secondary importance is the form through which that scope is conveyed. Distinctive form without magnitude of content will not yield a work of genius. For example in drama, rigorous observation of the three unities of time, place, and action will not be sufficient to produce a great play in the classic tradition unless the content is that of a Sophocles or a Racine. Likewise, a more loosely-knit dramatic structure will lack masterly effect unless the material is that of a Shakespeare or Marlowe. The same kind of observations apply to conventions in poetry and fiction; and, outside literature, to those in the visual arts and in music. The great artist, in whatever medium, makes exceptional use of form and convention through choosing exceptional material. This argument recalls what was said in the opening sentence of this essay.

Finally, let us return to another point made in the first part of this essay, and elsewhere in this book: the rarity of genius. Such rarity means that certain individuals are exceptional in relation to not only their own period but all periods. The truth of this will be clear from considering, in connection with the present day, any of the figures we have referred to in this essay. They would be giants in the contemporary context even as they were, whether fully recognised or not, in their own day. It must be assumed that this situation will continue.

Given this assumption, the conclusion follows that the many egalitarian tendencies in modern Western society should be solely directed at establishing equality of opportunity. Any aim beyond this, such as the creation of a higher frequency of genius-attainment than was found in the past, can only be regarded as unrealistic. While available resources should not be spared in establishing equal opportunities, the upshot of such equality should not be overestimated. This line of reasoning, in defending the principle of equality of access, is immune to charges that it endorses the glaring deprivation of access unfortunately found in past societies; while, at the same time, it advises against the excessive optimism which some people attach to the application of the principle.

Change & Continuity in Western Intellectual Culture

It perhaps goes without saying that Western intellectual culture is complex in a very wide range of ways. One extremely important aspect of its complexity is that it possesses an extensive historical heritage which is both valued and, at the same time, viewed critically. With this heritage, it has witnessed both change and continuity. As with all complex cultures possessing a heritage, the change and continuity have always been relative to each other, in the sense that concepts of change have depended, for their meaning, on concepts of continuity and tradition. Thus, alteration has had a clear meaning because tradition has had too. By contrast, a traditionless society which had always been subject to sudden changes would have only a very vague sense of what change actually was.

The West's tradition and heritage derives ultimately from Greek and Jewish sources: from the rationalistic, scientific, ethical, and artistic aspects of ancient Greek culture, and from the ethical and ontological aspects of Judaism and Christianity.

Reliance on the latter was the main feature of mediaeval culture (and of such culture as had existed before this in the Dark Ages). The main feature, but not the only one: Greek thought also found a place in mediaevalism, chiefly the ideas of Aristotle, but incorporated, as far as was logically possible, into the Christian perspective.

Then came the break from mediaevalism: the Renaissance, followed by the emergence of modern science and philosophy in the 17th century. These developments, in the wake of which we are still living, signified, in the long term, the spirit of ancient Greece: its open-minded rationality.

Overall, the cumulative inheritance is a living reality to all informed and thinking people, while at the same time being seen as a product of the past, and therefore as subject to the limitations of the periods and circumstances in which its various achievements were attained.

Returning to the break with mediaevalism: this requires emphasis. The advent of modern science and philosophy was to a considerable extent a new development in the West after the collapse of Graeco-Roman civilisation. One has to say 'to a considerable extent', because the development did not totally undermine, or even divest itself of, the influence of Judao-Christian ontology.[1] However, it did eventually modify that influence to an enormous degree; and its radical wing instigated the ontological conflict between science and religion which has since become a defining feature of Western culture.

The changes became so far-reaching that they established a new kind of ontological and epistemological movement, one totally different from that of mediaevalism because it was based chiefly on the open-minded, empirical, and scientific way of thinking previously referred to. The radical wing of this movement, in challenging the ontological claims

---

[1]   Ontology is the key issue here. Through all the intellectual changes which have taken place in the West, a number of the *ethical* ideas in Judaism and Christianity have retained their appeal, and remain a major aspect of the Western way of life. By and large, it is not these which have placed obstacles in the path of intellectual advance. Those obstacles have mainly come from the dogmas of religious ontology.

At the same time, as has been earlier noted, the intellectual advances of the 17th century provide instances of several leading thinkers seeing no conflict between their intellectual work and their religious beliefs. Descartes, Leibniz, Locke, and Newton were Christians, and Spinoza was a pantheist. But, with time, the situation changed substantially: see next part of main text.

of religion, gained the upper hand among most leading Western philosophers from the mid 19th century onward.[2]

However, this predominant radicalism has recently been confronted by a resurgence of religious belief, especially the fundamentalist kind, and not only from Judao-Christian sources but also from Islamic. This resurgence has presented major problems, yet stands little chance, in the long run, of ending the prevalence in the West of the radically empiricist spirit. The latter will probably not only retain the upper hand but be reinforced as a result of doing so. Its further enhancement will be welcomed by all who, in common with thinkers such as Comte and Durkheim, regard untrammeled scientific empiricism as the ultimate stage or form of human thought: ultimate not in the sense that, with its ascendancy, all objective knowledge is possessed, but in the sense that it is the only sure means of gaining such knowledge.

Again on the assumption of its continuance, what will in turn be maintained is a corpus of unprejudiced and non-dogmatic thought constituting an integral feature of Western society as — in modern times — a largely open society. This corpus represents intellectual culture of high order. Bearing in mind what has previously been said about the rarity of genius, the supposition must be that the most significant contributors to this culture will be (by definition) a minority, as they are now. In this respect too, continuity will persist.

In addition to these points about genius, it should be noted that there are also, as there have always been, many lower-level contributions to the culture; and that these have a complex gradation. This gradation is also likely to continue.

Given these considerations, let us turn briefly to the subject of economics. In the West, and to a substantial degree throughout the world, the economic norm is industrial capitalism, and increasingly monopoly capitalism. Industrial

---

[2]   If names need to be mentioned, then reference to Schopenhauer, Marx, Nietzsche, Spencer, Santayana, Russell, Dewey, Ayer, and Sartre will amply serve to illustrate this point.

capitalism replaced mercantilism as the latter had, in its turn, supplanted agrarian feudalism. Capitalism may well, in due course, be replaced by some form of socialism, or at least be radically modified to the extent of becoming far more subject to public monitoring and guidance than it has ever been before. Either way, the dominant position it has occupied in the West for the last 200 years may well come to an end.

However, even if change does occur, it is unlikely to impact decisively on high-order intellectual culture. This is not only because the latter's chief contributors will remain a minority but also—a related reason—because economic issues are not those that fully engage the energies of the most advanced minds. In a survey of the West's, indeed the world's, leading philosophical doctrines, it will be found that economics generally occupies only a small space.[3] This is not to argue that economics is not important as an intellectual discipline. For a host of obvious reasons, it unquestionably is. Nevertheless, its importance lies largely in the dimension of breadth—in its relevance to the ubiquitous human need for physical well-being and security—rather than in the dimension of depth, where more complex issues arise once the need for physical well-being and security has been met.[4] Philosophy's perennial themes[5]—the most intellectually demanding to be found in the ontological, epistemological, ethical, and aesthetic spheres—carry the mind far beyond what is required to secure basic physical flourishing.

---

[3]   Notable exceptions to this rule are the works of Marx, Engels and Spencer, and, to a lesser extent, those of Plato. But it must be stressed that the economic emphasis is not the norm in philosophy. It is of course the case that the majority of those primarily interested in economics are professional economists, whose range of thinking generally is not, and does not have to be, as panoramic in character as that of philosophers.

[4]   Compare this argument with the way Nietzsche defines 'high culture': "A high culture is a pyramid: it can stand only on a broad base; its very first prerequisite is a strongly and soundly consolidated mediocrity. The crafts, trade, agriculture… are in no way compatible with anything other than mediocrity in ability and desires" (*Twilight of the Idols*: trans. R.J. Hollingdale, London, Penguin Books Edition, 1968, pp. 178–9).

[5]   Some of these, it will be recalled, were referred to in the previous essay.

In brief conclusion: like all intellectual cultures of high order, the Western evinces a delicate balance between change and continuity. For example, the emergence of modern science and philosophy did, in due course, largely break with a previous tradition, that of mediaevalism, but harkened back to an even earlier one, that of ancient Greece. Even mediaevalism itself, though mainly a departure from Graeco-classical culture, was not completely so. Further, large-scale changes in economic system, though very important socially and politically, have left the recurrent issues in philosophy mainly unaffected (a fact which points to the limited character of the impact of economics on complex culture). In addition, these philosophical issues continue to be extensively addressed by a minority of people, in accordance with degree of intellectual ability. The same point about minorities applies to scientific advance and the many changes this brings about. It is of course true that change is undeniably change, but its linkage, in some way or other, to what has gone before is never broken.

Philosophy's Position in Contemporary Western Society

Among evolutionary biologists it is generally agreed that the brain, in sub-human animals and in man, developed initially as a survival mechanism. If this view is accepted, the assertion can be made that there is an original and fundamental linkage between mental operations and will-drives: drives to survive, and flourish biologically, and satisfy various physical needs and impulses. Thus intellect can be seen as originally an instrumental, and in this sense secondary, factor in human evolution, the primary one being the biological imperatives to which intellect was instrumental.

Given the long and complex history of human development since the emergence of intellect as a survival tool, the question to be considered is this: to what extent has intellect remained an instrument for the satisfaction of biological needs, and to what extent has it moved beyond this role? In tackling this question, social contexts must always be taken into account.

In answer to the question's first part: it is quite clear that much mental activity is still in the service of biological drives. The securing of shelter, sufficient food, physical well-being and comfort, plus the satisfaction of the sex-drive and the desire for children: these remain projects which are virtually universal, and which call for a large measure of intellectual energy. In all these activities, the mind evinces an intensely practical orientation, operating in a strictly purposive way to achieve a specific goal. In this regard, the mind is very much a medium for motives.

In answer to the second part of the question: manifestly, a great deal of intellectual effort, at least in complex Western societies, does extend much beyond biological satisfactions. A sizeable number of people have intellectual — and emotional — needs which can only be satisfied by the exercise of skills and expertise not necessary for simple physical well-being. (It perhaps goes without saying that these needs and skills are indices of advanced biological evolution.) The specialist abilities required by high-skill professions such as medicine, law, teaching, journalism, and engineering carry intellectual activity into much wider areas of experience, knowledge, and judgment than are covered in the pursuit of basic biological satisfaction. Clear-cut purposes and goals are also involved here, but obviously of more complex kinds, because, among other things, they entail a stronger relation between the individual and society as a whole. Through the professional activity, the individual is, or should be, contributing significantly to *society's* well-being, and not just to that of himself and his family. Further, the outstanding practitioners in the various professional fields are the people who possess not only the requisite capacities but also the most developed vocational sense of social contribution.

However, even these people are, in the main, less intellectually active than a small group whose engagement goes beyond the strict remit of almost all the paid professions, in that it is panoramic, inter-disciplinary, integrative, and often completely voluntary. This engagement can best be described as philosophical: chiefly an attempt to present a general view of the structures and processes that constitute total reality and/or an attempt to argue for a particular way of behaving. The first enterprise is ontological and epistemological, the second ethical. Both aim at maximal generality.[1]

---

[1]   For reasons of economy, space will not be given to the other main discipline in philosophy: aesthetics. But what will be said in connection

Philosophers have hailed from all sections of society; though, increasingly since the 19th century, they have been university lecturers; hence the main professional context has been academia. But, whatever the context, the chief point to bear in mind is that philosophers seek to be as inclusive and synoptic as possible in their thinking. This endeavour makes more demands on the intellect than any other because of its integrative range of reference to achievements in the various individual disciplines: the sciences and the humanities. At the same time, philosophy clearly cannot just consist of references. It must display perceptual originality, in relation to attainments in the sciences and humanities.

The scale of such thinking obviously places it at the furthest remove from that aimed solely at securing biological well-being. In this sense, it can be said that philosophy is the most advanced form of intellectual activity.

Again, as with all levels of intellection, purposes and motives are involved. In ethical philosophy, the aim is to influence behaviour. In ontology and epistemology, the aims are, to an unrivalled extent, those of enlightenment, clarific-ation, liberation from illusion. It is true that clarification has practical-utility value, but it has much more than this, and ontology and epistemology focus on the 'much more'. What ontology and epistemology offer is offered chiefly for its own sake, as an end in itself rather than as a means to an end. In this, they are wedded to the highest achievements in art and in pure science. Thus we can say that the finest products of intellect in the ontological and epistemological spheres transcend practical purposiveness in their ultimate effect, presenting a picture of things which is meant to induce, in James Joyce's phrase, "luminous stasis".[2] In this

---

with ontology, epistemology and ethics, about the general importance of philosophical thinking, will implicitly apply to this area too.

[2]   It is partly in this sense that Wittgenstein says that philosophy "leaves everything as it is". Also in this connection, let us recall Marx's famous saying that philosophers have only interpreted the world, whereas the point is to change it. Marx's implication is that the ethical (changing the world) can transcend the ontological (interpreting the world). The reply

way, intellect, having begun as a secondary factor in human development, has become a primary factor. This is a deeply enriching consideration.

Yet beside it must be posed the question: how many people will respond to ontology's and epistemology's offers, and, indeed, to the ethical aspects of philosophy as well? It has already been noted that practising philosophers are a minority of the population, most working in universities. Also, as far as the historical record shows, they have been a minority in the past. These two facts strongly suggest that, short of a complete revolution in cultural habits, philosophy will remain the activity of a few. Moreover, it can be assumed that philosophy's readership, also now a minority of the population, though a numerically larger one than in earlier periods, will remain such.

Additionally, we have noted that no more than a "sizeable number" of the population possess high-calibre professional expertise, and are therefore occupationally required to think at a level which, if not strictly philosophical, at least approaches the philosophical. Numbered among them are probably a few practising philosophers, and almost certainly most of philosophy's lay readership.

Outside the universities and the sphere of specialist professions lie, of course, the majority of the population, who are not occupationally obliged to think at levels approaching or including the philosophical. While certain individuals among the majority are philosophically active, most are not; and of the latter, a preponderant number show no sense whatsoever of being in need of engagement with philosophy. Given that the majority are under neither professional nor cultural pressure to be philosophically involved, it would seem that, barring the cultural revolution

---

to this is that the ethical can indeed bring change, but only to an extent, though this extent is considerable; what cannot be changed must therefore be left as it is, as an object of luminous stasis.

previously referred to, this non-involvement will continue into the foreseeable future.

The above statements should not be interpreted as an attack on the majority. They are simply observations of social facts. These facts may be partly or even largely the result of educational and cultural deprivation, and it is certainly the case that such deprivation is far less a factor at the social levels from which most university lecturers and professional specialists are drawn. However, whether deprivation is an explanatory factor, and, if so, the degree to which it is one, are questions that only empirical invest-igation can answer. Whatever the explication of the situation, the latter is as stated, and the point stands that a massive transformation in attitude and lifestyle will be required to change it.[3]

In the interests of balance, it needs to be added that one group which is not part of the majority, the leading figures in the worlds of industry, finance, and commerce, also, by and large, display a lack of philosophical engagement. They rarely show interest in ontological issues and, if they have any ethical sense of social contribution, usually confine it to points about providing employment and consumer goods (while seldom referring to their profits or their influence on government). Of all non-majority groups, they are the ones whose power in society is the most disproportionate to their general intellectual calibre.[4]

Let us now make some concluding remarks, including recapitulatory ones. Philosophical thought, like that in the arts and in pure science, transcends concerns about the simple biological well-being of the individual, and indeed all forms of self-interest (prominent among which is economic self-interest). If ontological and epistemological, its object is

---

[3]   This consideration should, incidentally, be borne in mind by those who choose to describe the majority as "the working class", and who advocate a "working class culture" based on the majority's present cultural situation.

[4]   These brief points echo a larger argument made in a previous essay, 'Anti-Bourgeois Attitudes in the West'.

to enlighten the social collective; if ethical, to secure a morally acceptable flourishing of the collective, of which the individual is of course part.

While every effort should be made to eradicate cultural deprivation and so increase the possible range of philosophical involvement, what should never be under-estimated, at all social levels, is the role which continues to be played by biological and economic will-drives. These drives stand in stark contradistinction to the kind of thought which transcends them, and from which alone genuine philosophy can emanate.[5] Neither to be underestimated are the envy or indifference felt by those of lesser intellectual capacity toward those of greater: towards those who are true philosophers.

The continuance of a large measure of thinking which is solely driven by biological or economic interests, and yet which often masquerades as something 'higher', constitutes an ongoing problem in contemporary Western society. So too does the perpetuation of envy or indifference toward thinking which *is* genuinely higher. Philosophy's position in contemporary society is, as said, far from central and secure. Its marginal status should be frankly acknowledged by its adherents, who must also recognise that future advances in economic well-being are no guarantee whatsoever that philosophy will, as a result of those advances, come to play a decisively larger role in the general consciousness. Previous economic advance has not produced this result, and there is no reason to assume it will in future.

---

[5]   For more on this crucial distinction, see Schopenhauer, whose writings on this subject remain a *locus classicus.*

The Aesthetic
& the Moral

Thomas Mann, in his novella *Tonio Kroger* (1903), has his central character say: "No problem, none in the world, is more tormenting than that of the artist and his human aspect." What he means is this: the artist, and especially the writer, is a person of exceptional powers of perception and penetration. As a result, his purview is far more extensive than that of the average person; consequently, he is able to comprehend the less aware (and less self-aware) people better than they are able to comprehend themselves. With this greater comprehension comes the power of depiction and representation: a capacity largely absent, or present in only very limited degree, in the average person. The possession of this ability carries, however, a number of implications which are morally dubious. The artist may come to regard other people as mere raw material for his creative work, simply a means to an artistic end — as, then, objects of aesthetic use.

For any artist with a genuinely moral sense, this attitude, while perhaps technically required, is nevertheless troubling — indeed, tormenting. It produces a sense of isolation and alienation from the generality of mankind, setting the artist apart in a way that clashes with the simpler, social and communal impulses which he also possesses as part of his human make-up. The artist, no matter how great, shares a number of tendencies with average humanity; yet the practising of his art denies him an outlet for the satisfaction of these tendencies.

In delineating the artist's situation, Mann is not saying anything essentially new; others before him (especially Henry James) and since have expressed more or less the same views. Yet Mann does, in this novella, invest what he has to say with a memorable succinctness and poignancy: so much so that the story has become, for many people, a *locus classicus* of the ideas it conveys.

Clearly, what is at issue here is a conflict between, on the one hand, highly developed intellectual and aesthetic capacities, and, on the other, the simpler aspects of human life and the felt need to act morally as well as artistically. No one, Mann and other artists are saying, can achieve full human satisfaction or fulfilment through art alone. While art is very important, it is not the only important thing. The artist needs to be a moral and social being as well as a creative one; and the greater the artist, the greater the need. This does not mean that the artist should ever compromise the clarity of his vision, but it does mean that this clarity should always coexist with questions about authentic social and moral relations with others. What is sought is not compromise or evasion but efforts at connection. "Only connect", famously urged E.M. Forster, who was, incidentally, a strong admirer of Mann. Also, while connection may not be feasible beyond a certain point, up to that point it should be fully engaged with. Obviously, there are cases where a large measure of communication is not actually possible, but whatever measure is possible should be fully realised.

It is noteworthy that most great writers have been deeply concerned with social and moral issues as well as aesthetic ones. Homer, Dante, Shakespeare, Milton, Goethe, Tolstoy: these are among the figures who, in their work, have made vital connections and achieved memorable syntheses. Returning to Mann: widely regarded as the greatest European novelist of the 20th century, he also attained a synthetic and capacious outlook as his career progressed. *Tonio Kroger* was a relatively early piece, published when he was only 28, and was succeeded by a series of works which, while continuing to focus on the person of exceptional mentality, explored a range of issues confronting modern man in general.[1] Indeed, even in *Tonio Kroger*, a concluding note is the importance, to the artist as to everyone else, of "that love of which it stands written that one may speak with the tongues of men and of angels, and yet having it not is as sounding brass and tinkling symbols". The Biblical quotation about the spirit of charity, a spirit which overrides aesthetic fastidiousness, has a universal dimension, and suggests that, even in this text, the author has already found at least the beginnings of the resolution of the conflict which he previously articulated so powerfully.

---

[1]   See the reference to Mann in the opening essay, 'Politics and Neo-Darwinism'.

An Inescapable Duality

At the base of the statue of St. Paul in the gardens behind St. Paul's Cathedral in London, there is an inscription which includes the words: "such scenes of good and evil as make up human affairs." If it is true that the terms 'good' and 'evil' still carry meaning in modern culture (as their continued popular usage suggests they do); if it is also true that this meaning is a precise one for religious believers and non-believers alike, and that there is, not complete, but considerable agreement[1] between the two groups about its meaning; then the above-quoted words convey an idea which is very important to grasp.

The idea is that actions deemed either morally positive or negative are a perennial feature of the human situation. They have been, are, and will remain components of all human experience. (It is in this sense that Henry James Sr. described contact with evil as the experiential norm.) Thus the human context displays an irreducible doubleness, an inescapable duality.

Generosity coexists with greed; benevolence with power-hunger; probity with perversity; altruism with selfishness. Mostly the divisions are between individual people, but sometimes they are actually within the same person.

---

[1] The point about 'considerable agreement' is crucial. For most non-believers, ideas of good and evil are creations of the human mind, not discoveries made by it. (See opening essay, 'Politics and Neo-Darwinism'.) For most believers, on the other hand, they are discoveries. While this difference makes complete agreement impossible, an important area of concurrence still remains, about a number of actions which are to be described as good or evil, and about what action to take against the latter.

In these observations there is, of course, nothing new. They have been recorded by humanity's great minds, both secular and religious, since ancient times. Some sense of the timespan involved here can be gauged from the considerations that, for example, Judaism and Zoroastrianism focused their doctrines, in the B.C. centuries, on the ubiquitous facts of good and evil behaviour; and that, in the 19th century A.D., the same was done by the atheist Schopenhauer and the Christian Dostoyevsky; likewise, in the 20th century, by the Christian T.S. Eliot and the atheist Sartre.

No society in recorded history has been devoid of evil, just as none has been of good. The same is of course true of the present-day world, where the gold-glow of altruism is bordered by the black of blood-soaked ruthlessness; where honesty struggles to maintain its place against fraudulence; and where modesty of material desires lives at the edge of the huge shadow of avarice. The perpetual character of mankind's dark capacities means that at no time can the future be realistically viewed in utopianist, or even near-utopianist, terms. This point was made with conspicuous force by Reinhold Niebuhr in 1945, at the end of World War Two, when in fact a utopianist way of thinking about the future was entertained by a sizeable number of people.[2] Regarding the task of building a world community on the ruins of war, this—says Niebuhr—will be "a perpetual problem", and cannot be approached realistically without an understanding that "moral ambiguities[3]… are permanent characteristics of man's historic experience". Hence, those assuming the task must always be "prepared for new

---

[2]   This attitude was understandable enough, given the horrors of WW2, and the consequent hope that they could be succeeded by a kind of heaven-on-earth.

[3]   These ambiguities were on full display in the so-called international settlement of 1945. In sealing the defeat of European and Japanese fascism in the name of humane values, Soviet despot Stalin negotiated with an established imperialist power, Britain, and a power which, many argue, was aspiring to a new kind of imperialism, the United States.

corruptions on the level of world community which would drive simpler idealists to despair".[a]

Over sixty years on, we, of the post-WW2 generations, can see exactly what Niebuhr meant, as will our successors, sixty years hence.

[a]   For all quotations, see *The Essential Reinhold Niebuhr*: ed. Robert
       McAfee Brown, New Haven and London, Yale University Press,
       1986, pp. 126, 128.

Scientific & Poetic Modes of Describing Physical Objects in the World

[This essay strongly reflects what was said in a previous one, 'Genius and Changes in Social Context', about the distinction between, on the one hand, imagistic and sensuous modes of description, and, on the other, modes which are sub-imagistic and sub-sensuous.]

Firstly, by 'physical objects in the world' is meant physical entities existing outside of, and independently of, the brain of the person making descriptive statements.

Secondly, strictly scientific statements about the above entities aim at complete literalness, at being totally free of figurative elements (metaphor, simile). By contrast, poetic statements about such entities abound in figurative language and are nourished by it. The more literal the form of statement, the less figurative and personally idiosyncratic it is; vice versa, the more poetic the form of statement.

In the scientific community, the intent is to link all descriptive statements together, in an effort to produce a coherent, unified, and systematic body of truth-claims. Attempts are continually made to establish relations between averrals. Clashes or contradictions between them are examined scrupulously, with the aim of overcoming discrepancies. Also, considerations are constantly made about research programmes that can be generated by averrals. The intention of these programmes is to yield further statements which can then extend the established system of factual claims.

The endeavour to produce a unified system of statements, a coherent corpus of doctrine, is what makes scientific statements publicly shareable in a way that poetic statements are not: shareable not only within the scientific community but, in varying degree, beyond it, among the general public. What is being shared is a joined-up way of thinking, in which all the parts are mutually supportive. It is only this kind of thinking that can produce an inclusive picture of the structures and processes which constitute physical objects. And it is only such a picture that can be a genuinely *common* property of a physically descriptive kind.

The differences between the above types of statement and poetic ones will be obvious. Poetic ways of describing objects are not, and do not aim to be, parts of a coherent system. While they are publicly shareable, they are definitely not so in the way described above. Each, with its personally idiosyncratic use of language, is powerful aesthetically, stirring the senses, imagination, and emotions of the reader; but it cannot become part of an interlinked schema which weds it to statements about the same object made by other poets. Each is *sui generis,* with its own independent context. A statement made by one poet on a particular object may be like that of another poet, but cannot corroborate the latter in the way two scientific statements can corroborate each other. Parallelism is not mutual validation. Poetic statements do not reinforce each other in any logical or empirical sense. Nor, at the same time, do they invalidate each other in these senses. While it is true that one poetic statement may be judged *aesthetically superior* to another, such superiority is not logical or empirical invalidation of the other.

Further, poetic statements are almost always made in imagistic, sensuous, impressionistic, or emotive forms. Such forms convey, often of course with great power, a *personal experience* of the object being described, and are therefore distinct from forms conveying a description which attempts to transcend personal experience. Science always aspires to achieve the latter description; which is precisely why it seeks to go beyond the personally idiosyncratic and experiential mode. While, as said, poetic statements are publicly shareable, they are so as *shared personal experiences,* and not as bodies of doctrine which aim to have a supra-experiential character.

It might be counter-argued that science actually does not succeed in going beyond experience in its endeavour to describe objects in a supra-experiential way. The experience in question is not just that of the individual human being but of the entire human species. This argument is essentially Kantian. Moreover, it is one which the scientific community must seriously consider, given the fact that the whole scientific project is anti-Kantian in the sense that its strict philosophical position is realism—the view that the world beyond the human mind can be accurately accessed, to at least some degree. This is in contrast to the Kantian position that the world beyond the human mind cannot be accurately and objectively accessed at all, but only experienced in a human-subjective manner.[1]

However, to repeat, it is the *objective* of science to gain accurate access, which means attempting to break free of subjective modes of experience. Whether or not it actually achieves this objective is an enormously complex issue: one where, at a philosophical level, the debate is essentially between realism and Kantianism. But engagement in this debate is outside the parameters of the present essay. Keeping within those parameters, suffice it to repeat the point which began this paragraph, and to add that poetic statements have no such objective, nor claim to have it.

The supra-experiential and systemisable character of scientific descriptions means that they are the only acceptable ones for inclusion in the effort to describe physical objects in a fully coherent manner. Yet poetic modes

---

[1]   More will be said on Kantianism in a later essay.

of description are perfectly acceptable in other contexts, to serve other purposes, where scientific coherence is not being aimed at. In fact, in these other contexts, the scientific mode of description often has little appeal or psychological effectiveness when compared to the poetic. This discrepancy is not actually a problem. Culture is complex, involving very different areas of interest and enthusiasm. Not everybody is intent on achieving the descriptive unity which science aspires to, and that is fair enough. Many people will always be less interested, for example, in a scientific description of first sunlight on a hill's horizon than in Shakespeare's rendering of it in the words: "the morn in russet mantle clad / Walks o'er the dew of yon high eastward hill",[2] and they are perfectly entitled to their preference. Linguistic usage can excel in many different kinds of ways, creating many different sorts of edifying perspectives for the reader; and one way—undoubtedly, as its perennial impact shows—is the poetic.

---

[2]   *Hamlet,* I, 1, ll.167–8.

The Recognition of Causality

The above phrase is of course an adaptation of the famous dictum that "freedom is the recognition of necessity". This definition of freedom is to be found in, for example, Spinoza and Hegel. Its implications are as follows: 1) To be ignorant of causality is to be mentally unfree; hence knowledge is mental freedom; 2) Knowledge, being freedom, is also practical effectiveness—effectiveness in the world is only possible through awareness of how nature works. This links with Bacon's sayings that only through obeying nature can we command it, and that knowledge is power.

What is common to both these implications is the view that causation is pervasive; that its omnipresence is what must be acknowledged. This view is of course the basis of philosophical determinism.

However, the determinist position, in its original and classical form, has been subject to modification in the light of three key considerations. Firstly, there is the philosophical argument, articulated fully by Hume, that causality is actually no more than a concept, a concept which is an interpretation of our experience of regular sequential conjunction between events.

Secondly, there is the development of quantum theory in modern physics. This appears to show that, at the microscopic level of the physical world—the level of individual atomic happenings—there is no causal connection between events.

Thirdly, again in connection with developments in modern physics, causal regularity or law, as experienced thus far, is regarded as statistical only: in other words, it will *only probably*, not certainly, be encountered in the future.

Thus the future is not bound to be like the past, since the causal laws assumed from our experience so far are not known to be invariant. Because they are not known to be this, they cannot be regarded as necessary, and therefore can only be viewed as statistical. Hence we are talking about the recognition of statistical causality, not of necessity.[a]

So, a modified determinism, as applicable at least to the physical world, is as follows: causality is the most reasonable interpretation of the fact of constant conjunction between events. Though causality's existence cannot be proven, it can be validly assumed. Causation, while apparently not existing at the microscopic level, can be assumed to obtain at the macroscopic level—that of large numbers or aggregates of atomic events. In fact, causality can be taken to be the cement of the physical world at the macroscopic level. There, it is unbudging and cannot be shaken off, just as a physical object cannot shake off its shadow. Thus any human action attempting to change situations in the physical world must inescapably engage with causality, and negotiate it just as a sailing vessel must negotiate the winds.

At the same time, however, we must always be ready to alter our formulations of causal law, should future experience require this. While accepting the omnipresence of causes, we should constantly be prepared to change our notions of causal pattern.

Now, returning to the 'human action' previously referred to, let us recall what was said, in the essay 'Genius and Changes in Social Context', about physicalism: the argument that all mental processes are really physical brain-processes. On this view, a decision to act in order to change situations in the physical world is itself a physical process, a brain event, and is therefore caused just as every physical event is. It is part of what is called the causal closure of the physical. Proponents of physicalism, regarding mental events as really physical ones, see all brain activity as part of the warp-and-woof of physical causality.

[a] This distinction between statistical causal law and necessary causal law requires me to acknowledge that, in my previous book, *Economic Reform and a Liberal Culture* (Exeter, Imprint Academic, 2010), I advanced an inconsistent argument. In one of the book's essays, 'Determinism and Prescription' (pp. 36–8), I cited with approval Hobbes' equation of causality with necessity, in the essay's opening quotation from *Leviathan*. Yet I went on to contend that causal laws are, in relation to the future, probabilistic only. This of course is another way of saying that they are statistical. So, I was incoherent in first accepting Hobbes' identification of causality with necessity, but then in characterising causality as statistical.

Mechanistic &
Teleological Causality

Determinism in the strictly scientific sense argues that events are caused one at a time, and that the individual event would be different if its cause(s) had been. Thus event-sequences consist of a series of contingencies, of point-by-point dependencies of effect on cause. This individuated view of causal relations is defined as mechanistic. It is the view that predominates in present-day science.

By contrast, there is the notion that causal relations are not individuated and contingent but holistic; that whole series of events, as distinct from individual ones, have to be what they are, in a specific kind of way. This contention posits a factor existing prior to the series which shapes the latter and pervades it. Usually, the postulated factor is a purpose, aim, or design of some kind.

The point about a postulated *a priori* factor must be stressed in the light of the argument that there is an inescapable sense in which *any* series has to be what it is, simply because the events comprising it have to be what they individually are. The latter argument is indeed valid; but it is a very different position from the one which postulates purpose and design. In the design argument, the series is not regarded as simply the sum total of all the events it contains, but as the resultant of the purpose or design which initiated the series as a whole. This view of causal relations is described as teleological.

Also, the alleged shaping factor is to be differentiated from the concept of causal law, as deployed in science. If it is agreed that causal laws are not actually *things* in the world but, rather, the thus-far-experienced regular behaviours of things in the world (the behaviours being the event-sequences); if it is also agreed that a category distinction must be drawn between a thing and its behaviour — then laws cannot be factors of the kind postulated by the teleological argument. A purpose, aim, or design is arguably a thing,[1] whereas a law is not a thing shaping behaviour but, as said, the behaviour itself.

The teleological position is frequently found in religious doctrines, where purpose or design is attributed, not surprisingly, to deity. But the position is also present in much historicist thinking, not always of the religious kind, where entire sequences of historical events are seen as following a pre-established design.

Though the teleological outlook has previously been described as non-scientific, there is one sense in which some people might forcefully argue that it is scientific. In biology, we have the phenomenon of organic potentiality being actualised in the process of physical growth and development. Is not potentiality — teleologists may contend — the shaping and pervasive factor which is absent from the mechanistic view?

In response to this question, the first point that needs to be made is the somewhat obvious one that biological potentiality is not a purpose, aim, or design in the mentalistic sense in which most religious believers view divine purpose. Nor, indeed, is it obviously mentalistic in any other sense. It is, arguably, something physical. Secondly, and again obviously, it is far too general a concept to apply to historicist doctrines, whether religious or not. There is no mileage whatsoever in saying that the whole of human history can be understood solely as the unfolding of a set of potentialities established (either by deity or by some other cause) at a particular point in man's physical evolution. Clearly, considerations of potentiality apply as much to the many *pre*-historic stages of human development as to the historic.

---

[1]   If mental states are re-described as physical brain-states, then a purpose can be defined as a thing in the sense of being a specific brain-state.

Thirdly, a precise definition of biological potentiality needs to be given. As previously contended, potential is something physical, not mental. Hence, it is not a bodily entity in its own right, separate from other bodily things, but is the *entire physical structure* of the nascent organism. In popular parlance, the organism is said to be 'endowed' with potential, but it would be more appropriate to say that the organism is its own potential: its capacity for development is its bodily make-up.

Seen in this way, potentiality can be regarded as the mechanistic effect of previous physical causes (largely genetic ones). Further, it is an effect with determinate features and characteristics which make certain lines of development possible but which preclude others. These determinate qualities are the specific components of the mechanistic effect.

As regards the process of development after the formation of potentiality, attention should be drawn to the terms 'genotype', meaning what an organism has the capacity to become, and 'phenotype', meaning what that organism actually does become in a particular and constraining environment. All actual development is phenotypical, since all organisms are of course located in particular environments. Now phenotypical growth is very much a matter of point-by-point dependencies of effect on cause, of individuated causal relations. It is, then, very much a matter of mechanism.

This said, the obvious point must be made that an organism, if given a suitable environment, will certainly realise elementary potentialities, even if it fails to realise non-elementary ones. A baby horse will grow into an adult one; and a baby human will, anatomically at least, become an adult human. Given these facts, the teleologist may still assert that these processes are more than mechanistic, more than an accumulation of individuated sets of causal relations. He may claim that they are holistic, and collectively driven by the force of initial potentiality. However, if, as previously argued, potentiality is defined as physical structure, then what is being claimed by the teleologist who accepts this definition[2] is that structure is the sole agency producing events, rather than being an entity which instigates development only in conjunction with environment and external circumstances. It is as the latter that structure is best defined. Potentiality, even the most elementary kind, is activated only by interaction with environment. Developmental dynamic resides solely in that interplay, and never just in processes within the organism.

These points sit well with the mechanistic viewpoint. Previously, structure was described as a mechanistic effect, and the determinate features of that structure as the specific components of that effect. What can be added now is that these components create the specific conditions for the ways in which the organism interacts with environment. Given such conditions, only certain interactions can occur. Conditions, then, constitute limits, just as they constitute positive possibilities within those limits. From these conditions result dependencies, mechanistically, and, through all the interactive processes which may transpire, these dependencies can never be overcome. Thus contingent factors are mechanistically derived, and pervasive. Mechanism, then, is what is at work in the actualisation of biological potential. This, finally, is the more overtly the case the more extensive the interactive process, therefore the more numerous the point-by-point contingencies as the organism negotiates environment with the inner capacities it possesses.

To conclude: mechanism can viably be regarded as universal in the physical world. Hence teleological doctrines connected with religion, in so far as these relate to the physical sphere, must be judged invalid. So too must those connected with history and biology. The physical world contains only individuated sets of causal relations, each

---

[2] Of course, the teleologist may not accept it. But, if so, he is obliged to offer and justify another definition.

contingent only to the preceding set, and none the resultant of any pre-established design covering all sets. Hence the physical sphere cannot be understood through reference to any argument from design.

For mechanists, it is by now perhaps commonplace to say that mankind inhabits a physical universe that is both cause-pervaded and undesigned; and that the fact of event-regularities, otherwise called causal laws, does not signify the existence of cosmic purpose of any kind. If these are indeed commonplaces, then possibly it is less routine to say that they make, on men and women, a much more exacting demand than would otherwise be the case, to show courage, resolution, and constructiveness of an unflagging kind.

Isaiah Berlin on Determinism: A Reply

[This text is reproduced by kind permission of the Editor of the *Ethical Record*.]

[For background to this essay, the reader may wish to consult not only the two preceding essays but also the following material on determinism which I have previously published: the essay 'Some Arguments for Determinism', in *Progressive Secular Society* (Exeter, Imprint Academic, 2008); and the essays 'Determinism and Prescription', 'Compatibilist Freedom and Global Causation', and 'Some Further Arguments for Determinism', in *Economic Reform and a Liberal Culture*.]

[Further, while in the present book I have viewed determinism mainly with reference to physical brain-events, this essay defends determinism as a general doctrine, and therefore can be read sympathetically by proponents of mental causation.]

This essay examines a series of points on the subject of determinism made by Berlin in his book *Four Essays on Liberty* (Oxford, Oxford University Press, 1969). These points are found in the Introduction (pp. ix–xxxvii) and in the essay 'Historical Inevitability' (pp. 41–117). Because the Introduction is comparatively brief, I will not give the page numbers when I refer to particular parts of it and quote, but I will do so when dealing with 'Historical Inevitability', because of its length. Also, while the long essay is primarily an attack on notions of historical determinism,[1] it does contain negative criticism of all doctrines of continuous cause and effect, especially scientific ones; and this criticism is what I will engage with when I do make specific references.

Berlin's arguments are significant not only because of their individual content but also because, taken together, they constitute the standard and traditional argument questioning the validity of the determinist outlook. Note, 'questioning' the validity, as distinct from flatly denying it; Berlin does not claim that determinism is a false doctrine, only that it is a doubtful one, and "I know of no conclusive argument in favour of determinism". The point about the absence of conclusive arguments is clearly a cogent one. No intelligent determinist would contend that there is absolute proof of continuous causation, just as, one assumes, no libertarian (believer in the existence of human freedom from causality) would argue that such proof exists for his position. The issue, as always in philosophy, is a matter of the relative strengths of arguments. Berlin, then, goes no further than casting doubt on determinism. However, it is very extensive doubt, and is cast in, as said, the traditional mode. This mode is what will be subjected to criticism.

Berlin's questioning approach is part of a line of thinking which, as he himself emphasises, stretches back to Kant. Essentially, the point is this: if causality is pervasive, as determinism argues, then moral concepts, at least in their traditional forms, are seriously undermined, if not rendered totally meaningless. He writes:

> when Kant said that if the laws that governed the phenomena of the external world turned out to govern all there was, then morality—in his sense—was annihilated...
> In a causally determined system, the notions of free choice

---

[1] I concur with Berlin in rejecting all forms of historicism as causalist doctrines. See the previous essay, and later essay 'Morality and Progressive Historicism'.

and moral responsibility, in their usual sense, would vanish,
or at least lack application.

The "moral responsibility" of which Berlin speaks is equated
by him with freedom in the libertarian sense. This is
indicated by his coupling of the phrase with "free choice";
and by, in another part of the Introduction, his referring
sympathetically to the traditional view of "ordinary men"
that people are morally responsible for their actions "since
they could have behaved differently".

He argues that to remove the postulate of libertarian
freedom from moral discourse "takes the life out of a whole
range of moral expressions"; and he goes on to contend that
"very few defenders of determinism have addressed
themselves to the question of what this range embraces
and… what the effect of its elimination from our thought
and language would be". The effect, he insists, would be
"upsetting". This point about the perturbing and unsettling
consequences, on our moral thought and language, of
accepting determinism is repeated later in the book (pp. 70,
72, 76). His position is, in essence, an *argumentum ad
consequentiam.*

To be especially noted in Berlin's reasoning are, firstly,
endorsement of an established outlook—partly, it seems,
because it *is* established; secondly, the view that any
challenge to the outlook could only have a negative and
destructive outcome. Further to be observed is that,
crucially, *no empirical argument* is offered to support the
contention that libertarian freedom does actually exist.
Support for the contention hinges, not on empirical or
scientific considerations, but on ones to do with language-
usage, long-standing ways of thinking, and age-old general
attitudes. This kind of approach is inappropriate to what is,
strictly speaking, an ontological and scientific issue: either
libertarian freedom exists, at least in some measure (Berlin's
own qualification to the postulate), or it does not. Con-
versely, either causality is pervasive, or it is not.

When, elsewhere in the Introduction, Berlin does look at the issue in scientific terms, the results are in fact more fruitful. He fully accepts the role of science in modern culture, while at the same time, and quite reasonably, questioning whether the scientific outlook, based as it is on assumptions of universal causation, can cover the whole of reality: "to proclaim that science is the search for causes… is not to say that all events have them." This is unquestionably an important point, one to which determinists must give serious attention. However, the determinist does point to the ever-increasing explanatory range of the sciences — explanation of course being about cause and effect — and therefore to the decreasing areas of reality which remain unaccounted for in terms of causal process. Moreover, he argues that there is no reason to think that, given the requisite resources, the scope of scientific explications will not go on expanding, with no conceivable end in sight. The coverage may, of course, never be complete; but science's ultimate capacity or incapacity for completeness is something that can never be known in advance.

Further in connection with science's explicatory scope, another point made by Berlin needs to be noted. He argues that the success of scientific methods in illuminating the non-human areas of reality does not automatically mean that such methods will be successful in dealing with the human area. Again, this is a significant claim. Nonetheless, it has been found that explanations of a physico-chemical kind which apply to the non-human area apply also, in considerable measure, to the human. This is obviously the case in medical science. Additional evidence for linkage between the non-human and the human is found in the fields of evolutionary biology and astronomy; as regards astronomy, there is, for example, the fact that the physical constituents of the human body are the same as those found in the rest of the physical universe, even in the most distant stars.

Inevitably, in due recognition of science's achievements, Berlin concedes that knowledge "of scientifically established laws" renders us "more effective" in dealing with the world. He also acknowledges that:

> there is plenty of empirical evidence for the view that the frontiers of free [libertarian] choice are a good deal narrower than many men have in the past supposed, and perhaps still erroneously believe.

This statement is strongly echoed later in the book, where reference to causal processes is actually extensive. Berlin says that science has shown that the scope of human choice is

> a good deal more limited than we used to suppose… that human beings *more often than not* [italics mine] act as they do because of characteristics due to heredity or physical or social environment or education… it is salutary to be reminded of the narrowness of the field within which we can claim to be free.

He goes on to say:

> and some would claim that such knowledge is still increasing, and the field still contracting. Where the frontier between freedom and causal laws is to be determined is a crucial practical issue. (All three quotations from pp. 73–4)

Nevertheless, despite the concessions to science, we see the persistence of the assertion that there is an area which lies, and indeed must lie, outside scientific coverage. It persists, apparently, as an *a priori* assumption, and one deemed inviolate. The position is not *a posteriori* with any definiteness, not clearly the result of empirical investigation. This point is bound up with a certain vagueness in the first of the three quoted statements. Has the "empirical evidence" thrown light on causal processes *only*, or on these *and* non-causal areas of reality? If only the former, then there are no empirical grounds for saying that the latter exist at all; if what has been empirically ascertained is only areas of causality, then no justification is available for saying that non-causal areas exist as well.

Likewise, with the second quotation, nothing is cited as empirically confirming the existence of libertarian freedom. The "knowledge" referred to is only that of causality. And what systematic line of enquiry, other than the empirical-scientific, could confirm freedom's existence? Berlin himself says: "What can and cannot be done by particular agents in specific circumstances is *an empirical question* [italics mine], properly settled, like all such questions, by an appeal to experience" (p. 71). Again, then, what empirical method can show that freedom is a reality?

Furthermore, relating these points to what Berlin has previously said about moral responsibility, precisely *which* actions are free and therefore to be judged in traditional moral terms? According to what he says in the second quotation, only infrequent actions are. But how does this square with the 'ordinary man's view', previously cited with approval, that actions in general are free, and therefore could have all been replaced, at the time of their occurrence, by other actions?

Berlin's inadequacy in justifying the postulate of freedom's existence is again evident when he contends that, unless causal processes

> are held to leave some freedom of choice—and not only of action that is not determined by choices that are themselves wholly determined by antecedent causes—we shall have to reconstruct our view of reality accordingly; and this task is far more formidable than determinists tend to assume.

This, clearly, is familiar ground: unless the postulate of free choice is made, a host of stressful thought-consequences follow. Yet again, no empirical grounds for making the contention are given. What we are given instead are what might be called socio-consequentialist grounds: think in a certain way in order to avoid the thorny social and cultural problems which arise from thinking in a different way.

Berlin's repetition of the non-empirical argument suggests that the chief thrust of this kind of reasoning is to establish a logical compatibility between the various moral concepts it uses. The objective, then, seems to be to harmonise the meanings of moral terms. This is evident later in the book (p. 65) when Berlin cites Kant's view that "where there is no freedom there is no obligation; where there is no independence of causes there is no responsibility and therefore no desert". This statement certainly hangs together as a mutually supportive use of terminology; but, once more, nothing is done to show that "freedom", meaning "independence of causes", actually exists. Without such demonstration, the meanings here assigned to the terms "obligation", "responsibility", and "desert", cannot be accepted unquestioningly.

So far, no response has been made to Berlin's claims about the alleged intellectual shortcomings of determinists. Now, here is the response. To reiterate his words: "very few defenders of determinism have addressed themselves to the question of… this task [which] is far more formidable than determinists tend to assume." Actually, far more than a few determinists *have* confronted the issues to which Berlin refers. They are fully cognisant of the complex and challenging implications of their doctrine. They realise that acceptance of their views would indeed entail an extensive change in moral discourse: there would be transformation, even upheaval. Moreover, the stress these developments would undoubtedly produce is seen as no argument against them.

In this regard, let us return to the traditional notion of moral responsibility. Determinism does use the term 'responsibility', but in what is, by comparison with libertarianism, a reduced way. Determinists openly acknowledge the different usage, and make no attempt to obscure the point. They regard a person as responsible for his action in the restricted senses that he is the one who performed it, and is therefore the author not only of it but of its consequences.[a]

Also, his responsibility includes his endorsing the action when he regards the latter as morally valid. This endorsement can be expressed in such words as: 'I am glad I am being carried along by my own inner desire to do what I regard as morally defensible.'

At the same time, this view of responsibility does not of course extend to the contentions that he performed the act by exercising libertarian freedom, and that, therefore, he could have acted otherwise on the occasion in question. (In passing, it can be added that feeling remorse for an action and its consequences does not depend on the view that the action was uncaused. An action may be seen as caused and yet still be regretted. Indeed, the regret too can be regarded as caused.)

Now, according to Berlin and traditional thinking, the conviction that the agent possessed libertarian freedom, and therefore could have acted differently on the said occasion, provides the justification for apportioning, on the one hand, censure and blame, and, on the other, praise and congratulation. Further, this conviction is what gives these terms their actual content and meaning. Without it, the terms would, as earlier quoted, 'have the life taken out of them'.

Moreover—and this is to his credit—Berlin effectively anticipates the determinist's reply that such terms would not be rendered useless by acceptance of a doctrine of continuous causation. They could, as determinists do in fact argue, be used as incentives to producing morally desirable behaviour in the future. Censure and blame could be applied with the intention of reforming the agent's behaviour for the future; and praise and congratulation, with the aim of maintaining the agent's past good behaviour. In all cases, the agents would be treated with a view to the *effects* which the treatment was meant to produce. Thus the treatment would be fruitful within the orbit of deterministic thinking.

Nevertheless, as Berlin quite rightly says, censure and praise are given *not only* in situations such as these. They are applied also to people in the past, now dead and so beyond the possibility of change or persistence in behaviour. In these cases, censure or praise have a validity quite separate from that of the moral strategy outlined above. According to Berlin, this validity can rest only on the view that the people in question exercised libertarian freedom in performing the actions for which we retrospectively appraise them. In other words, even if the determinist argument can be accepted in the case of prospective moral strategy, it cannot be accepted in the case of retrospective appraisal.

This is a powerful point, but it can be met by determinism. Determinists argue that praise or censure of people now dead does *not necessarily* depend on the view that they acted with libertarian freedom, just as praise or censure of those now alive does not. It contends that, in all cases, past and present, causality is operative, and that this claim would be confirmed if scientific investigation could be fully applied in all cases. What, then, are the common grounds for praising or blaming both the living and the dead?

In answering this question, the determinist first comments on certain aspects of the psychology of praise and blame. Focusing on psychological issues is an appropriate procedure because this in fact is what Berlin himself does when he refers to and defends traditional attitudes, customary ways of thinking and feeling, in the moral sphere. Determinism notes that, foremost in the minds of many people when they congratulate or censure, is *the action itself*, and the quality of moral goodness or badness which they either confer on the action or regard as intrinsic to it. The same applies to a consideration of the agent's motives, and of the consequences of the action.

Those people who think this way voice approval or disapproval of actions in a very specific and focused manner. Their attitude is so concentrated on making such point-by-point judgments that they do not look beyond the complex of action-consequence, and do *not* ask if the action was free in the libertarian sense. They do not say: 'I am praising or censuring this action only because I regard it as uncaused. If I thought it was caused, my attitude would be entirely different.' Indeed, if they do think at all analytically about their attitude, they are more likely to think in causalist terms: 'a good action arose from a good motive, a bad action from a bad motive.' They may even go a step further and aver: 'This man's motives were good because he has/had a good character, while that man's motives were bad because he has/had a bad character.' In essence, they are voicing a spontaneous warming to, or aversion from, a set of events.

To repeat, these are observations about the *psychology* which many people display when they praise or blame. The observations are not meant to show that determinism is correct. Though, of course, determinists do think their doctrine is true, their aim in making these observations is not to demonstrate this correctness but to convey the point that it is not psychologically inevitable and universal for praise and blame to rest on belief in libertarian freedom. Yet Berlin and the traditional school hold that it invariably does: "men have, at all times, taken freedom of choice for granted in their ordinary [moral] discourse." This, clearly, is too selective a view.

Because Berlin's thinking hangs on the assumption that libertarian freedom exists, he focuses on the issue of "deserts", and argues that, unless people are regarded as free, they cannot be said to deserve punishment for their misdeeds. For Berlin and all traditionalists, the issue of deserts is inextricably linked with that of justice: justice is the meting out of deserts. He goes on to say that, in determinist discourse, the concept of deserts has no place, or at least no place of the kind it has customarily occupied in moral reasoning.

On this latter point, he is absolutely correct. However, his accuracy does not weaken the determinist position. Determinists eschew the concept of deserts, which is a *retributive* one, in favour of the concept of *reform*. If the wrongdoer, on the occasion of his wrongdoing, could not have acted otherwise, then, argues determinism, he should be subsequently treated in a way which exclusively aims at improving his conduct in the future, provided that improvement is actually possible. This is the moral strategy referred to previously. As was also specified about this strategy, reformative thinking is deterministic because causalist effects are being sought, and effects are the outcome of events. Further, even if the intended effect is not achieved, then whatever outcome does emerge will also be the effect of causes. The causalist character of the thinking means that, if reformation is attained, then the agent will be caused to behave in a morally acceptable way, just as he was previously caused to behave in an unacceptable way.

To those who reply that this doctrine amounts to regarding people as automata, the determinist repeats another point made earlier: that many people, when considering an action of which they morally approve, do not stipulate, as a condition of approval, that the action had to be uncaused. The thought that the action was or may have been caused does not bother them; they are perfectly content to place value on the action itself, and even on its motive. Hence they will readily approve of an action resulting from a reformative process. Indeed, they are only too pleased to see a genuinely good action performed, in a world where such a phenomenon is all too rare. This attitude, though often spontaneous and not the product of systematic thinking, is clearly consonant with determinism as a fully thought-out philosophical position.

Furthermore, for those who accept the postulate of causal continuity, there are manifestly different kinds of evaluation to be placed on different kinds of continuity. A continuity leading to generous or heroic action will be seen to require a more positive epithet than 'automaton-like'. The latter terms imply lack of dignity, significance, distinctiveness, poignancy. They are obviously inappropriate in a wide number of contexts.

An additional riposte from the determinist is that a sharp distinction must be drawn between, on the one hand, having a negative and fixed attitude toward the postulate of continuous causation, and, on the other, being prepared to examine the postulate in a completely empirical and open-minded way. People may not *like* the proposition, but of course a mere attitudinal stance such as this is scientifically worthless. All that counts scientifically is willingness to investigate; and if investigation reveals continuous causation, then scientifically the matter is settled.

Returning now to the reformative process: determinists argue that this must often involve tough and harsh measures. The intention is to make the wrongdoer fully aware of the destructive character of his action, and of how much society disapproves of it. The malefactor needs to understand that, while other people regard his misdeed as caused, they also think that *it would have been better*, from a general and supra-personal moral standpoint, if the action had not been perpetrated. The production of such awareness in the wrongdoer, hopefully resulting in a genuine change of heart and therefore of behaviour, is seen as itself resulting, in part at least, from severe treatment. The degree of severity depends, of course, on the nature of the misdeed. Also, to counterbalance the harshness, other reformative practices are recommended: for example, education, counselling, and other ways of helping the malefactor develop a more positive attitude to society. However, some measure of severity always remains part of the causalist argument.

Actually, the determinist's recommendations are, as practical measures, more or less the same as those made by libertarians. The only difference is that the determinist sees such measures as aimed at producing effects rather than the absolutely free choices which libertarians think in terms of. Since the determinist has a concept of a reformative *process*, he has a sense of interlinked events, since this latter phrase defines what a process is.

A further point needing to be made about reform — and it is one that opponents of determinism frequently neglect — is that the determinist places great emphasis on the *capacity* to reform. Everything that has been said so far about how behaviour can be improved assumes, of course, the ability to improve. Without the latter, treatment of the wrongdoer cannot be reformative at all, only preventive. (It is clearly the case that reformative treatment is frequently combined with preventive; but here we mean treatment which is exclusively the latter.)

The stress placed by determinists on capacity means that they are not guilty of a simplistic charge often levelled against them. This charge is as follows: Berlin discusses what he regards as the determinist argument that people are caused to act in a certain way because of their 'character', and because they are "made that way". However, the determinist's concern with capacity, as a dynamic and developmental factor in behaviour, means that he does not argue that a caused action on a particular occasion will be inevitably duplicated on all subsequent occasions, and by virtue of the person's 'character'. While it is true that the term 'character' is part of determinist discourse, the term is a complex and flexible one. It denotes, among other things, a set of potentialities, only some of which may be actualised in specific actions on specific occasions; indeed, some may never be actualised on any occasion. Thus, if a person can be said to be 'made' in a certain way, can be said to have a certain constitution, that description designates only his capacities; and where the latter involve the ability to alter and improve behaviour, then the person can be said to be 'made' in a way which enables reform.

At the same time, determinists do insist that a person's set of capacities is finite,[2] and that this finiteness is itself caused (chiefly by hereditary factors). However, the point about finiteness is a completely separate one from the above

---

[2]   For more on the group of points given in this part of the essay, see the next essay: 'Essence, Existence, and Identity'.

argument that, within the parameters of a fixed number of abilities, considerable plasticity of behaviour is possible.

One more point about Berlin's reasoning needs to be made before we draw to a conclusion. He briefly touches on what is actually a vexed issue for libertarians: to claim that a person's actions are uncaused, by pre-existing motives, intentions, or any other antecedent factors, is to claim that they are random; and "is not random behaviour the very opposite of… rationality, responsibility?" Indeed it is, replies the determinist. For libertarians, freedom must involve rationality and responsibility in order to be morally defensible; and here Berlin is questioning whether random freedom can possess moral status. This query is part of a broader line of thinking in which he concedes that libertarianism faces a very difficult problem: behaviour which is random can no more be called morally responsible than can, according to libertarianism, behaviour which is caused. Hence, reasons Berlin, "a new model" of thinking, one not yet devised, is required to save the notion of moral responsibility and therefore that of libertarian freedom. This model would transcend the traditional antithetical concepts of caused action and random action.

Note that the new model is needed exclusively in order to preserve the twin concepts of responsibility and liber-tarian freedom. As has already been said, determinism assigns a reduced (though still substantial) role to the concept of responsibility; and, of course, no role at all to that of libertarian freedom. Hence determinism does not need the new model which Berlin hankers after. Further, the deter-minist doubts whether such a model is in fact possible, either logically or empirically. How can actions be neither caused nor uncaused? What further context for action can there conceivably be? Berlin grants that the creation of the new model "needs a philosophical imagination of the first order". One only hopes that, by the word "imagination", he means the ability to make imaginative leaps in order to return to reality and achieve a firmer grasp of it, rather than the ability to establish merely imaginary constructs.

Now, in conclusion: reference was made earlier to the continually widening circle of causalist explanation in science. This expanding process has not been confined to the non-human areas of reality but has increasingly extended into the human sphere, especially in the fields of psychology and jurisprudence. In the latter particularly, almost all of the progressive changes which have taken place in legal thinking in the West in modern times have been influenced by considerations of causality. These changes have led to a much more humane and comprehensive approach to criminality than was even remotely the case in past, pre-scientific eras, when belief in libertarian freedom held virtually undisputed sway. There is, of course, much progress still to be made in this sphere, where the notion of libertarian freedom still occupies a position. However, that this position is now greatly reduced is, for the determinist, highly significant in the ways already specified.

To echo points made earlier, there is no reason to think that causalist ways of thinking will not continue to spread beyond the non-human sphere, to all areas of consideration about what actually transpires in the world. While Berlin is right to say that no assumption should be made that causalist thinking will eventually be found to be universally valid, the opposite assumption should not be made either. In addition, since the area of causalist explanation is increasing, and that of non-explanation decreasing, it can be argued that, more and more, the burden of proof lies with those people who contend that there are parts of reality which causalist explanations will never reach. The history of science shows that this contention has been repeatedly refuted across the centuries.

The determinist, of course, has nothing to fear from the expansion of explication. Further, he stresses the point that explication is not part of the libertarian picture of free action.[3] Something else which holds no fear for the determinist is the radical overhaul in moral theory which, as Berlin accurately observes, is entailed by acceptance of determinism. Totally jettisoning the twin notions of retributive justice (i.e. deserts) and libertarian freedom would indeed constitute an upheaval, even in the present-day legal context where, as said, a good deal of deterministic thinking has already been taken on board, producing extensive transformation. Focusing entirely on either reformative or preventive treatment of the wrongdoer, or on a mixture of both, would be concomitant parts of this reorientation. The latter is seen by the determinist as inescapable, given the power of scientific and philosophical arguments.

Human thought does and should evolve. Like everything else on the planet, it is subject to change. For the determinist, requisite change in thought is true progress.

[a]   *Cf.* interestingly, the libertarian Sartre's definition of the word "responsibility" as "consciousness of being the incontestable author of an event or of an object". See *This Is My Philosophy*: ed. Whit Burnett, London, George Allen and Unwin Ltd., 1959 (1958), p. 212.

---

[3]   It perhaps goes without saying that reference to effects should be no part of libertarian discourse. Yet one wonders if there is such a reference in a particular libertarian argument, sometimes heard, that a person should be fully informed of the harsh treatment which awaits him if he commits a crime, and should be left in no doubt about the legal consequences of his action, *so that* he will then be in a position to make a completely enlightened and absolutely free choice about whether to perpetrate the crime. Surely the state of mental enlightenment is something aimed at as an *effect* of providing him with information. It may be that, as libertarians argue, what follows from this mental state is not an effect; but still, it seems, at least one effect has been incurred.

Essence, Existence, & Identity

The three previous essays on the subject of causality have affirmed a determinist position. This position always needs to vindicate itself in the face of libertarian arguments. One of the most cogent assertions of libertarianism in recent philosophy has been that of Sartre, who presents a number of arguments which the determinist is required to answer effectively. Sartre's libertarianism is bound up with his doctrine of atheistic existentialism. According to this doctrine, there is no such thing as a human nature or essence, if defined as a factor in human beings which precedes and conditions their actions. There is no such factor solely because there is no God to create it. It could exist only by divine fiat, which the non-existence of deity of course renders impossible. Since, then, human beings do not possess any essence prior to their actions, whatever essence they may be said to possess can only emanate from their actions. The latter, crucially, are free in the libertarian sense. Further, if people's actions are defined as their existence — hence the term *exist*entialism — then existence is prior to essence, not the other way round. At the same time, however, the essence arising from their actions is never a fixed or final entity, because action is ongoing, continually transpiring; so, no once-and-for-all, determinate essence is ever attainable.

Now, in critical comment on the above: the determinist may well concur with Sartre's atheism, but, in doing so, he is not logically obliged to agree that there is no factor within human beings which conditions their actions. His position is that there is indeed such a factor, and that this is *potentiality*. Potentiality or capacity, as has been previously argued, determines the possible parameters of all behaviour, and so plays a conditioning role. Given this role, it can be defined as essence: thus essence, as capacity, is condition for action.[1] This means that essence is prior to existence if, in accordance with Sartre's own reasoning, existence is defined as action.

Also, over and above capacity, there are further factors prior to action, but these are ones which activate capacity as distinct from constituting it. Action is possible only when the potentiality for it exists and when that potentiality is stimulated.

However, having said this, the determinist still finds Sartre's arguments appealing in certain ways. Sartre's emphasis on the continuingly transpiring character of action remains important. Although action is conditional upon the existence of capacity, and is caused by factors stimulating the latter, still it gives the human being an *attained identity* which is not the same as an essence. A person is recognisable by, and appreciated through, his or her actions in a much stronger sense than s/he is through the possession of capacity for such action. And this identity is continuingly extended through ongoing action, the ongoing realisation of capacity. Thus, while essence precedes existence, existence precedes attained identity. The latter is conditional upon existence, which is in turn conditional upon essence.

The emphasis on identity as an effect or resultant of actions is important because it connects with the view that effects have their own intrinsic status and significance, ones quite distinct from those of their causes. Achieved identity

---

[1]   One may actually go further in discussing condition for action. In certain statements, a clear distinction exists between the condition for an action and the cause of it. E.g. 'The condition for my raising my arm is the ability of all my arm muscles to function properly, while the cause of my raising it is my overwhelming desire to pick an apple from an overhead branch'. However, if none of my arm muscles is functioning properly, and I try but fail to raise my arm, then that initial state of affairs can be regarded as the *cause* of my failure, and not just the condition for it. This point applies widely to situations of non-capacity, regarded as conditions or initial states of affairs. Thus the condition of inability to perform an action is the cause of failure to perform it.

occupies a space in the world quite separate from that occupied by its antecedents.

At the same time, mindful of Sartre, we must repeat that this identity is not fixed and final; or — and here we add to Sartre — is not fixed or final *while the individual still has potentiality to realise.* The point about capacity is recurrent and pivotal. Because potentiality has only finite parameters, the identity arising from its actualisation can have only finite parameters too. Identity remains non-final while there is more capacity to realise, but becomes final when there is no more to realise.

Returning again to Sartre: he rightly criticises those people who falsely regard themselves as having attained, at a particular point in time, a final state of being, in accordance with which they can exist in a fixed and static manner. But he is right to criticise them, and we are too, only when such people have not actualised their full potential. A state of underachievement is one to which further identity can, and should, be added. Consciously to persist in such a state, repressing thoughts about future efforts which one knows, or at least suspects, one is capable of, is to remain in a condition which Sartre himself aptly describes as self-deception or bad faith. Nonetheless, if a state were reached when all potentiality had been realised to the full, then such a state genuinely would be a fixed and final one, and beyond criticism. That such a state is feasible in the terms just specified is a notion which Sartre, with his libertarianism and his failure to equate essence with potential, erroneously rules out of court.

Emotivism: The Irreducibility of Feeling-Positions

[Despite what has been said in previous essays regarding physicalism, this essay will use mentalistic language in presenting its argument. It will do so solely for reasons of linguistic convenience.]

Hume says: "morality is determined by sentiment",[a] and "Reason is, and ought only to be, the slave of the passions."[b] In asserting the primacy[1] of affects, he stands as a central point of reference for all those who regard emotions and feeling-positions as primary and foundational in ethical outlooks. This position can be broadly described as emotivism. After Hume, writing in the 18th century, one of its leading proponents in the 19th century was Schopenhauer, to be followed in the 20th century by Ayer and C.L. Stevenson.

Emotivism, in arguing for the foundational role of feeling-positions, always needs to supplement this point by adding that these positions are communal and shared. Morality is social and collective in character; therefore its basis must be too. Clearly, a moral system can never be a mere assemblage of different individual viewpoints, each being applicable only to the individual who holds it. Such an assemblage would not be a system at all, only an arena of

---

[1]   This primacy relates to ontological commitment as well as to ethical. If love of truth can be regarded as the main motive for the ontological quest, then reason is clearly the servant of passion.

internal divergences and conflicts. A moral system, by definition, must unify individuals and establish some kind of homogeneity. While deeply respecting certain kinds of individual difference, it must assert definite behavioural standards to which all its members should adhere. As an ethical system, emotivism argues that these necessarily collective standards have their basis in feeling.

To echo points made in the essay 'Politics and Neo-Darwinism', these standards are inter-subjective or group-subjective (as distinct from being individual-subjective); and, being characterised by subjectivity, do not claim to be forms of moral knowledge. Their non-objectivist and therefore non-cognitivist status means that they are unrelated to theistic ethics. The latter invariably claim that the basis of morality *is* objective and cognitive, in the following ways: it is objective in the sense that it is decreed by the divine will,[2] and is cognitive in the sense that the divine will can be discovered by human beings. In stark contrast to this whole line of reasoning, emotivism contends that morality is a matter of feeling-based decisions and processes of justification which are entirely human in origin and reference.

The basis of this outlook being feeling, it is of course tautologically true that feeling is its irreducible element: what is foundational has nothing lying beneath it, nothing supporting it. However, more needs to be said on irreducibility. If it is argued that beneath feeling lies ratiocination of some kind, the question must be asked: what kind of ratiocination? Bearing in mind that we are in a non-cognitivist framework, the answer obviously cannot be: rationalising of a kind which leads to discovery of moral right and wrong—the kind that arises from having a series of empirical experiences which lead to the uncovering of a fact never before known. Nor can the answer be ratio-

---

[2]   However, what is not clear in these doctrines is whether objectivity attaches to simply being decreed by the divine will or to containing certain qualities which cause it to be thus decreed.

cination of a purely logical *a priori* kind, one not based on empirical experience but also leading to discovery of right and wrong.

If no further kind of reasoning can be invoked in the attempt to answer this question, then the assertion can be made that in no way is discovery involved in discourse on right and wrong. If it were, we would have morality of an absolute kind, because referenced to indisputable knowledge in the way science is or seeks to be. Without such referencing, morality must be regarded as relative; and, for the emotivist, that of course means relative to feelings.

At the same time, while rejecting the view that ratiocination is primary in ethics, the emotivist readily accepts the argument that it possesses secondary status; and that this status is in fact a very important one. Reason and the deployment of factual knowledge are needed for finding the best practical and technical means for achieving the ends which are themselves not the product of reason but of feeling. This finding of means is an extensive and complex process, and has increasingly involved recourse to the whole field of scientific discovery, as that field has hugely expanded in modern times. Reason's secondary status is, then, massive in scale.

Yet secondary it remains. It can never replace feeling in the formation of ends which subsequently seek means. Ends take shape as a result of an interaction between different streams of feeling and impulses, an interaction out of which a resolving and synthesising stream emerges, bearing a moral end or objective. It is true that this objective must then be subject to rational criteria, to check whether it is practically realisable, but this does not mean that it *emanates* from reason. It need only be compatible with reason and fact, after having been formed by sub-rational forces.

Returning now to morality's relative character, in connection with what has been said about the advance of science in modern times: the increasing separation of moral considerations from theistic religion has in large part been due to the growth of science, with its undermining of most of theism's ontological claims. This separation is a major phenomenon in modern Western culture. With it has come the view that mankind is inescapably thrown back on its own resources for vindication of whatever moral positions it adopts. Ethics, then, becomes entirely relative to human qualities; and, for the emotivist, the most fundamental of these qualities is feeling.

For a number of people, the view that vindication of morality cannot come from outside humanity — from a divine or other kind of transcendental sphere — gives rise to anguish, and this too has been an important phenomenon in modern culture (recorded with especial poignancy in the literature of existentialism). Though such anguish can be at least partly understood in terms of the continuing but perhaps only semi-conscious influence of traditional religious thinking, it nevertheless remains an acute exper- ience for some people, and these include some emotivists. It is not to be dismissed as an inadequate or atavistic state of mind; and emotivists not affected by the problem should nonetheless be sensitively aware of it. At the same time, it does need to be regarded as one which every effort should be made to supersede, however painful the process.

Emotivists as much as everyone else should be fully cognisant of the many problems attached to the vision of mankind's standing alone, ethically as well as physically, in a universe which, as far as present knowledge shows, contains no kind of moral consciousness other than the human.

[a]　Enquiry Concerning the Principles of Morals, Appendix 1.
[b]　A Treatise on Human Nature, III.

Kantianism & ...

Pragmatism

At first glance, there would seem to be nothing in common between these two doctrines. Kantianism, despite arguing that human beings cannot accede to objective knowledge of the world which exists outside the human mind, nevertheless asserts, as truth-claims in an absolute sense, that there *are* objective truths about that external world. It asserts, indeed, that there are objective truths about things in general. For example: there is a world external to the mind; the mind contains certain in-built attributes; the latter condition the way the mind experiences the external world, and so prevent the mind from gaining direct access to that world. All these contentions are absolute in the sense that no further criteria are applied to them in order to justify advancing them. The claims are entirely self-standing and unconditional.

However, this is not the case with truth-claims made by pragmatists. For pragmatism, a truth-claim is valid *only if* adherence to it leads to practical effectiveness in the world. Thus a condition is attached to postulating it: the practical consequences of accepting it. Because it is not self-standing, the pragmatic truth-claim is not of an absolute but of a relative kind. Indeed, pragmatism argues that truth in any absolute sense does not exist.

Since, then, Kantianism and pragmatism differ diametrically in the kinds of truth-claim they make, do they have anything in common? The answer is 'yes', because both doctrines are, in varying measure, philosophies of experience. Though Kantianism does make absolute truth-claims, it does so partly to focus on the centrality for human beings of their *experience* of the (objectively unknowable) external world. It argues that the only thing mankind can be absolutely certain of is its mode of experiencing the surrounding reality; and, equally important, that this mode is and can be its only guide for living in the world. Hence successful living depends not on possession of absolute truths about external reality but on effective navigation of our experience of that reality.

Pragmatism says essentially the same thing: meaningful living requires efficient negotiation of experience, not a command of absolute truths. At the same time, it adds a point which Kantianism does not make: to repeat, that absolute truths are non-existent. So, along with similarity, there is a difference. But the latter does not cancel out the former.

Neo-Darwinism

Kant's epistemological considerations about the structure of human mentality are highly relevant to neo-Darwinism: the latter doctrine focuses on the biological factors and forces which, in the process of evolution, have shaped human mentality, and have therefore shaped the relation of that mentality to the external world. At the same time, Kant's epistemological principles, if strictly adhered to, would make it impossible to affirm neo-Darwinism and evolutionary doctrine. For the latter, in the process of considering the relation of mind to environment, make knowledge-claims about the world external to the mind — and, according to Kant, such claims are inadmissible, since for him the external world is unknowable.

The Ethical Will

In the field of ethics, Kant's most famous dictum is probably, "Act only on that maxim which you can at the same time *will* to become a universal law [italics mine]". This is his 'categorical imperative', asserting that we should only ever act in ways which we would recommend for everyone else. In the quotation, the role of the word "will" is central. It implies, as said, an urging; and an arguing for, a contending and insisting. At the same time, it does not imply emotion. Kant explicitly denied that morality could be based on feeling, since, in his view, ethical action had to have a consistent and regular character, whereas emotions lacked consistency. Also, they could not be summoned up at will. Only an ethical will, a fixity of principle, possessed the required regularity; and this will was, for Kant, the sense of duty.

In this short essay, I will not be examining the ethical will from the emotivist standpoint explored in the previous essay. I wish simply to take note of it as a concept which weds willing to morality, and to distinguish it from the central idea on ethics found in the work of a later philosopher who was extensively influenced by Kant: Schopenhauer.

Schopenhauer, despite being decisively affected by Kant's work in epistemology and ontology, dissented from him in the ethical sphere. For him, emotion *was* the basis of morality. The emotion in question was, essentially, love of mankind—'menschenliebe'. Willing, in the Kantian sense, was not its basis. Indeed, for Schopenhauer, willing, in the Kantian or any other sense, was actually *im*moral, was the source of all pain and suffering; and morality was about overcoming willing of every kind, in favour of the intellectual and contemplative virtues, those leading to disinterested cognition, and to disinterested action to relieve suffering.

However, Schopenhauer can be criticised for failing to consider that actions arising from 'menschenliebe' might, and might have to, involve an act of will—even several such acts—for their implementation. He underestimated, therefore, the pertinence of willing to ethics.

At the same time, while the above cannot be said of Kant, many philosophers have regarded Kant's concept of will-based duty as being too rigid. Kant declares at one point that a person always has a duty to tell the truth, regardless of circumstances: "To tell a falsehood to a murderer who asked us whether our friend, of whom he was in pursuit, had not taken refuge in our house, would be a crime."[a] In reply, one might well argue that ethical principles, with their concomitant acts of will, need to be much more flexible than this, more capable of adapting to different situations, and (unlike Kant's position) concerned about the consequences of actions.

Thus, while Kant can be defended for linking will with ethics, the particular linkage he forges can be seen as too narrow, and insufficiently sensitive to the complexities of experience and situation. The role of willing in morality, a role which involves courage, tenacity and perseverance, is one which should entail deep and wide responsiveness to the multifariousness of problems and contexts, variegated as the latter are in the kinds of pain to be relieved, the needs to be met, and the aspirations to be fulfilled. Such responsiveness is informed, of course, not only by will but also by emotions of tenderness: hence the relevance of Schopenhauer's thinking on this subject. Will and feeling are both integral to ethical response, and this integration should mean the avoidance of rigidity.

[a]   As quoted in the article on Kant in *A Dictionary of Philosophy*: London, Pan Books Ltd., 1984 (1979), p. 192.

The Idea of Power &
the Secular Sense of
Awe

The sense of awe—that of the deepest possible admiration and reverence—has traditionally been most frequently manifested in theistic religion, the chief object of awe being deity. Bound up with this sense has been the idea of power: the power of deity, usually meaning deity's infinite capacity. The power-aspect of deity remains very strong in the world's three main monotheistic religions: Judaism, Christianity, and Islam.[1] Especially notable is the wording of the Lord's Prayer in Christianity: "For thine is the kingdom, the *power* and the glory [italics mine]."

However, as the growth of secular humanism has increasingly shown, it is possible to experience a sense of awe outside the framework of theistic religion. In the secular context, this experience is completely compatible with the possession of scientific knowledge and the causalist inform-ation this provides; since, the more that is known, the greater the appreciation of the complexity of reality. Also, that complexity is conveyed more by science than it ever was or could be by theistic ontologies, which are all pre-scientific in origin. That science and wonder not only do not clash but actually sustain each other is a point explored extensively in much secular/humanist literature.[2]

---

[1]   In passing, it is interesting to note that polytheistic religions, with the exception of Hinduism, do not now have global prominence.

[2]   A finely representative example of this is Richard Dawkins' *Unweaving the Rainbow*.

For the secularist, the complexity of the universe is generated by power, but that power does not reside in deity. It resides in natural processes, especially those which have been operative on this planet, and have produced biological evolution from unicellular organisms to human genius.

The secular concept of power differs from the theistic not only in the sense that this potency is located in a natural world devoid of deity, but also in the sense that it is without intentions or purposes. For example, the whole evolutionary process has been without a goal; the staggering variety that has emerged from it has been unintended. Thus this power has a certain purely spectacular quality, and it is partly toward this quality that awe is felt; as it is felt when watching the sun rising above a mountain peak, or giant clouds changing shape, or a mass of buffalo inching their way across a distant plain. The satisfaction lies in the sheer encounter with images.

A third way in which this notion of power differs from the theistic is that it possesses behavioural regularities at the macroscopic level of physical events, and therefore is not arbitrary in the way it functions; or, at least, it has not been arbitrary thus far. Its regularities make it, to a large extent, predictable. Hence it has so far shown itself to contain nothing of the grandiose randomness attaching to the kind of divine power in which many theists believe. Being regular and to a significant degree predictable, it is, arguably, limited. In other words, there may well be certain things lying outside its regularities which it cannot do; there may be certain things that cannot happen in the universe. If so, then the infinite capacity usually ascribed to deity by theists is not this power's possession.

Overall, the secular outlook, shaped by science, has a clear sense of the natural potency on which it focuses. Future scientific discovery will, in all likelihood, only sharpen that clarity, not replace it with some other perspective. Further, that this potency is held in awe is a point we need to return to, because it is one about attitude and feeling-position.

Bearing in mind what was said in the previous essay about the irreducibility of feeling-positions in ethics, we can say that, for the secularist, nothing lies deeper than emotional allegiance to the natural order in the way s/he thinks, feels, and acts day by day. Valued first and foremost as natural phenomena are all instances of human excellence. Valued next is that pervasive order of nature which has made such excellence possible and which, on a more general level, provides the only available conditions for human existence and flourishing. Additional positive evaluations are made, none owing anything whatsoever to super-naturalistic doctrines.

The superseding of the super-naturalistic, a process which can be regarded as perhaps the most important intellectual development of the last two centuries, finds definitive expression in the words of Santayana:

> When the heart is bent on truth, when prudence and the love of prosperity dominate the will, science must insensibly supplant divination, and reverence be transferred from traditional sanctities to the naked *power at work in nature* [italics mine], sanctioning worldly wisdom and hygienic virtue rather than the maxims of zealots or the dreams of saints. God then becomes a poetic symbol of the material tenderness and the paternal strictness of this wonderful world.[a]

[a]   In *The Wisdom of Santayana*: London, Peter Owen Ltd., 1964, pp. 256–7.

The Uses of the Term 'Materialism'

[This essay should be read in the light of the references to materialism made in the earlier essay, 'French Rationality in the 18th Century'.]

Let's begin by looking at the word 'matter', from which the term 'materialism' is derived. In modern Western philosophy, a key definition of matter, formulated by Descartes, is that it is an entity existing not only in time but also in space; and, being spatio-temporal, is 'extended'. As such, Descartes continues, it differs from 'mind', which exists only in time, not in space, and is therefore 'non-extended'.

This definition has provided the context in which, since Descartes, the word materialism has been used *in the most philosophically precise way.* It is true that, with Einstein and relativity theory, Descartes' classical distinction between time and space has been superseded, and that the latter are now seen as interconnected, so that whatever exists in time exists also in space, and vice versa. Hence the specific basis on which Descartes differentiated between mind and matter has been overthrown. However, the category differentiation itself remains closely associated with the work Descartes did on the subject; and there is no more fundamental meaning attached to the word 'materialism' than that which distinguishes the material from the mental sphere. Thus *philosophical* materialism is about matter's difference from mind.

Within the discourse of philosophical materialism there are two main positions. One is that reality consists of nothing but matter. This is eliminative materialism because, simply, it eliminates the mental from its picture of the world. The other, then, is non-eliminative materialism, which does regard the mental as part of reality, but as a part dependent on, because the product of, the material. The mental, then, is not eliminated but is ascribed a secondary status in the order of things.

One version of the second position is dialectical materialism, which argues that matter is not something static on which change and development have to be imposed, but something that contains within itself powers which provide the motive force for change.

Dialectical materialists contend that their perspective differs from mechanistic materialism for the above reason, that it sees matter as possessing internal dynamic. However, many mechanistic materialists also hold a dynamic view of matter, one compatible with the general mechanistic argument for the unbroken continuity of cause and effect. Also, incidentally, many mechanistic materialists are in the non-eliminative camp.

Outside the sphere of philosophical materialism, there are two other uses of the term 'materialism'. One is 'historical materialism': the view that economic factors are always the chief force shaping the formation of, and alteration of, social and cultural systems in their broadest features. Here, 'materialism' refers to economic activities and relations, and is therefore not to do with the distinction between matter and mind.

The doctrine of historical materialism is held by Marxists, who also adhere to dialectical materialism. However, it is perfectly possible to be an historical materialist without being one in the dialectical sense. There is no logically necessary relation between the two positions. Likewise, there is no necessary relation between being a philosophical materialist and holding either of these two views.

The final sense in which 'materialism' is used is the furthest removed from the philosophical. Here, the meaning is preoccupation with economic affluence and the acquisition of consumer goods. Thus a 'materialistic' society is one given to this preoccupation.

As all the above points imply, great care must be taken in the use of the term we have been discussing. The philosophical, historical, and consumerist usages should never be confused with each other, especially in the mass media, which are so influential linguistically.

Schopenhauer & Sartre

Born 117 years apart (1788, 1905), these two philosophers contrast and compare with each other in a number of very interesting ways. As is often the case when two thinkers are viewed side by side, the existence of stark differences in outlook between them seems, at first glance, to rule out the possibility of any similarities or parallels. Yet the latter are, on closer inspection, found to obtain. Such is the case with, for example, Mill and Nietzsche,[1] and with a number of other pairings which could be adduced; likewise with the two figures now under consideration.

Let's begin by examining the differences between them. Regarding Sartre, the reader will recall, from the earlier essay, 'Essence, Existence, and Identity', Sartre's insistence that a person can only be defined in relation to, and subsequent to, his actions; that this definition-through-action constitutes the person's essence; and that, since his actions constitute his existence, the latter is prior to his essence. In diametric opposition to this stands Schopenhauer's contention that a person's actions are a consequence of his character/temperament, and that the latter is fully formed before action transpires. Thus, if the word 'essence' is again used to designate what a person can be accurately described as, and the word 'existence' again designates what a person does, then for Schopenhauer essence is prior to existence — not the other way round.[2] Schopenhauer adopts the Latin

---

1   See, incidentally, my brief comparison of these two thinkers in *Economic Reform and a Liberal Culture*, pp. 94–8.

2   Thus, despite the similarities with Sartre which will later be discussed, there is no sense in which Schopenhauer can validly be seen as precursor

dictum *Operari sequitur esse* — what we do follows from what we are — to succinctly convey his position.

From this contrast emerges a further one. As said in the earlier essay on Sartre, he was a libertarian. From what has been said on Schopenhauer, he was clearly the opposite: a determinist, believing that a person's actions are caused by his character. On the other hand, Sartre, with his libertarian argument that our actions are caused by nothing, did not deploy the term 'character' at all; instead, as we have seen, he used the term 'essence', but even then only to designate how a person can be described with reference to actions performed thus far.

In addition to these differences, there are the facts that Sartre came to ally himself with the political Left, and was a political activist, whereas Schopenhauer was politically conservative, while at the same time not at all active in politics. Though giving much less space to politics than did Sartre, he wrote trenchantly on the subject, and was especially influenced by Hobbes. The distance between him and Sartre can be gauged from the consideration that the chief influence on Sartre's political thinking was Marx.

The gap between the two men is, then, very wide. Nevertheless, bridges can be built across it; let us now examine the similarities. Firstly, both thinkers were atheists. This common factor is actually more important than might first appear. Though, by the first third of the 20th century, when Sartre began publishing, atheism was not a particularly unusual position for a Western philosopher to take, it was highly unusual when Schopenhauer began publishing in the early 19th century. Schopenhauer was in fact one of the first Western philosophers since ancient times to be overtly atheistic; and so, in a broad sense, he helped pave the way for subsequent atheistic thinking, a general context of which Sartre was to become a part. More specifically, he influenced Nietzsche, who in turn influenced Sartre.

---

of Sartrean-type *exist*entialism. Given what that term meant for Sartre, this description simply cannot apply to Schopenhauer.

Bound up with this shared atheism was a deep involvement in the argument that theological ethics had ceased to be tenable in moral discourse. Both Schopenhauer and Sartre contended that, if there is no God, then there are no *a priori* moral values to be discovered in some sphere beyond the human intellect: a sphere such as a divine mind, or some other area of reality external to human consciousness. This meant, for both thinkers, that moral philosophy of a non-religious kind had entered a period of fundamental uncertainty and therefore crisis, since moral positions and actions could find no supra-human form of justification or vindication. The complete absence of the latter is a theme especially prominent in Sartre, in his exploration of moral anguish.

Here, what is particularly arresting is the historical dimension. Schopenhauer was speaking of moral crisis around the middle of the 19th century, and Sartre around the middle of the 20th, with very little difference in the content of the argument. That the crisis continues, in much the same form, in the early years of the 21st century, is an additional point to note.

Partly in connection with the crisis-issue, but also on a more general moral level, both philosophers extensively critiqued the ethical ideas of Kant. Schopenhauer saw Kant's concept of moral law as being indirectly derived from theological modes of thinking, and therefore something which had to be superseded by an ethical system in no way reliant on theism. Sartre too challenged Kant's notion of moral law, and initially regarded the Kantian categorical imperative of treating people as ends and not means as being totally unattainable. However, he later altered his view, treating this moral objective as indeed attainable, though still not possessing the status of law. In this, it should be added, his position was actually the same as Schopenhauer's. While rejecting legalistic notions, Schopenhauer advocated total disinterestedness on the part of the moral agent, which means treating people only as ends. At the same time, it should be stressed that both thinkers, in their moral advocacy, always refrained from laying claim to the kind of absolute certainty displayed by those whose ethical position is law-based or deity-based.

Finally, again in relation to their atheism, both philosophers viewed the sub-human natural world with considerable unease. Indeed, Schopenhauer was nothing less than horrified by the natural sphere, with its incessantly internecine struggle and turmoil. He argued that it could not possibly be the creation of a benign and omnipotent deity (as postulated by, for example, Christianity); nor could it be the actual substance of such a deity (as posited by pantheism). He saw the sub-human world, and in fact the whole of the cosmos beyond man, as being entirely without moral meaning or significance. This, in the 20th century, was to become Sartre's position as well, and that of those who shared his atheistic outlook. Such is the case even though Sartre did not describe the natural world in anything like the massive detail found in Schopenhauer.

Spinoza vs. Kant

In the earlier essay, 'Isaiah Berlin on Determinism', reference was made to Kant's view that moral concepts were meaningful only if human conduct was not subject to causation. Thus, causally determined behaviour could not be moral behaviour; the latter, by definition, had to be free in the libertarian sense. Ethics was irreconcilable with causality.

However, over a century before Kant, one of the most important books on ethics in Western philosophy was written by a determinist: Spinoza. For him, ethics *was* consonant with a causally-based outlook. In other words, ethical discourse, which is chiefly one of prescription, advocacy, and recommendation rather than of description,[a] was reconcilable with the view that behaviour is always caused.

The latter point needs to be emphasised as a key feature of Spinoza's outlook. He avers:

> There is in no mind absolute or free will, but the mind is determined for willing this or that by a cause which is determined in its turn by another cause, and this one again by another, and so on to infinity.[1]

Also, Spinoza's thoroughly deterministic position entails the compatibilist view of human freedom, that a free act is not one which is uncaused but one which is caused only by the wishes of the actor. In this sense, freedom is compatible with causation. As Spinoza says: "That thing is said to be FREE which... is determined in its actions by itself alone."[2] This

---

[1]   In *The Ethics and On the Correction of the Understanding*: trans. Andrew Boyle, London and New York, Everyman's Library, 1970 (1677), p. 74. Incidentally, this statement stands very close to one made by Hobbes in *Leviathan*: London and New York, Everyman's Library, 1970 (1651), p. 11.

[2]   *Ibid.*, p. 2.

free state contrasts with a thing which is not free but "necessary or rather compelled" because "determined in its... actions by something else"[3] external to it, e.g. extraneous coercion or force.

More specifically, a free and moral act is one caused by the actor's desire to preserve and enhance his/her own selfhood — and, importantly, the selfhoods of others.

It is highly significant that compatibilism is a position which Spinoza shares with a number of other leading philosophers of the past, such as Hobbes, Leibniz, Locke, and Hume; and with a growing number of recent and contemporary thinkers in the field of moral philosophy. This significance is bound up with the development in the West of modern science, a development dating from the 17th century — the century of all the above-named figures[4] except Hume. As hardly needs repeating, the scientific outlook at the macroscopic level is a causalist one; scientific explanation is constituted by reference to cause and effect. The wider the scientific mindset grows, the greater the concern with causation. Of this general tendency, Spinoza stands at the origin.

Spinoza saw no conflict between a concern with causation and one with ethics. In this, as we have seen, he differs from Kant. Kant insisted, for a variety of reasons, on separating science from morality. But in the present-day Western world, as science continues to expand, with no conceivable limits in sight, the question looms: is the Kantian separation legitimate and acceptable? Those who answer 'no' will find a staunch ally in Spinoza.

[a] For more on the distinction between prescription and description in ethical discourse, see, *inter alia*, my essay 'Determinism and Prescription', in *Economic Reform and a Liberal Culture*, pp. 36–8.

---

[3] *Ibid.*, p. 2.
[4] And, we might add, the century of Bacon.

Morality & the Doctrine of Progressive Historicism

Let's begin by defining what is meant by the above doctrine. This firstly requires us to enlarge on a previous reference to the word 'historicism', where it was characterised as the view that historical events unfold according to a pre-established design. Historicism argues that human history is shaped by laws which are specifically historical; and that these are independent of individual persons or groups of people, since the latter are located at particular points in time, whereas the laws transcend time-contexts. Every action of every person or social collective is caused by the operation of laws which determine the agent's behaviour. Thus every particular phenomenon in history is an effect of the workings of general forces.

Given this overall definition of historicism, the progressive version of the doctrine can be taken to mean that the operation of historical laws is continually developmental and constructive, unflaggingly leading toward better things, unfailingly progressive. This is the case even if progress is, or appears to be, slow, gradual, and painful rather than swift and easy.

Also, it should be noted that the progressivist perspective sharply differentiates itself from other historicisms: for example, from the view that history is cyclical, with event-sequences either repeating themselves in identical fashion or leading to equivalents and parallels of themselves; and from the view that history is a process of decadence, a continual decline from an initial (alleged) 'Golden Age'.

Clearly, the progressivist position sees history in very positive moral terms. If advance is essentially moral advance, then history has been the theatre of such development. And this, it must be emphasised, is history in its totality. Progressivism is not the view that history shows occasional but wondrous periods of moral gain, whose marvellous quality more than makes up for their scarcity, and provides an adequate sense of human advance. Nor is it a relativistic doctrine, according to which periods of moral flourishing are in many respects *sui generis*, self-validating in ways independent of each other—therefore not parts of an integrated continuum, not preludes or stepping stones to what succeeds them. On the contrary, it is an absolutist doctrine, in that it sees *everything* in history as part of a single continuum. Everything interlinks in a momentum which is generally progressive. Thus losses and setbacks, though they of course occur, are never decisive; they never ultimately deflect the course of history from its law-bound and upward path. Historical law is a law of amelioration.

Now, in comment on this position: two problems immediately suggest themselves. The first is that the whole notion of historical laws, whether meliorist or otherwise, is an extremely dubious one. It faces a number of objections not faced by the postulate of laws of nature, those of chemistry and physics, with which it should never be conflated. One objection is that, while natural laws are now, in line with modern physics, regarded as statistical only, laws of history seem to be regarded by their postulators as invariant, and this without any scientific justification. The suspicion arises that the postulate of historical law is based on an outmoded, Newtonian view of law.

Secondly, in the ethical as distinct from the ontological sphere, progressivism has to confront the obvious counter-argument that much of human history has been bloody and bleak, and has seen the frequent triumph of unscrupulous brute force and cunning. This point also applies to the present-day world, which will of course be history to future generations.

The observation about the role of force and violence leads us to ask the question: on what grounds does the progressivist see history in morally positive terms? Unless he adheres to a might-is-right ethos, he will have to regard many features of history as immoral and therefore regressive. This must surely be the attitude of anyone who holds what can be called a democratic-humanitarian outlook. Further, even if the progressivist accepts the use of violence to achieve democratic-humanitarian ends, he will still have to acknowledge that an enormous amount of past violence did not have this objective, even if some of its perpetrators were under the illusion that it did. He will have to recognise, in addition, that the past was a formidably complex mixture of elements: positive and negative, morally sound and perverse, constructive and destructive. Finally, he will have to accept the possibility, perhaps the probability, that this situation will continue.

On 'progress': definite criteria must be offered to define how the word is being used, since it is a value-term not a fact-term. Provided criteria are given, substantial arguments can indeed be made that there has been progress in history. One such argument is that history, at least in the West, has witnessed the gradual growth in the freedom of the individual and a concomitant decrease in group-authority. Two others, again applying to the West, are that political systems have become democratic, and that public welfare provision has hugely expanded. These developments are of course seen as progress by those who value individuality, democratic procedure, and general well-being. However, whatever values are at issue, an argument for progress can validly be made without reliance on the doctrine of progressive historicism: in other words, without dependence on the view that the progress achieved is a law of history.

If it is agreed that: all progress at all times has been the work of particular people in particular contexts; that it has been entirely dependent on those particularities and local factors, and would not have occurred if the latter had been different; that each set of particularities was what it was only as a result of, or in relation to, a previous set, and not as a result of any historical forces or factors pervading all sets; *then* it will also be agreed that there are in fact *no* laws of history, only *circumstantial contingencies* – and a staggering variety of them. Moreover, to return to an earlier point, these contingencies have often been of a very mixed character, and may well remain so. Hence the advances emanating from them are by no means guaranteed to last. In the much-quoted words of H.A.L. Fisher, "Progress is not a law of history. The ground gained by one generation can be lost by the next".

Over the last 100 years, historicist doctrines of all kinds, and not just the progressivist, have been in decline in the West. This has especially been the case since 1945, with the end of World War Two, arguably the most destructive and cataclysmic event in all history. In this period and throughout the 20th century, several of the West's leading minds were conspicuously engaged in an attack on historicism, particularly the progressivist kind. They included Russell, Santayana, Popper, Berlin, Heidegger, Ayer, Foucault, and Derrida. (This combined effort, incidentally, more than offsets the historicist efforts of Spengler and Toynbee.)

By contrast, the 19th century was much more an era of historicist thinking, chiefly of the progressivist kind, with Comte, Marx, and – most of all – Hegel as leading figures. (There is perhaps some connection between this phenomenon and the fact that the 19th century view of scientific law was still largely Newtonian.) On Hegel in particular: he, perhaps more than anyone else in the 19th century, sought to read moral (and rational) meaning into the whole of history. In this way, he strove to justify the course of history in its entirety — or, at least, the course which he *claimed* history had taken. Overall, the strongly progressivist element in 19th century thinking became a challenge which the most perceptive 20th century minds saw they had to meet.

At the same time, the 19th century also witnessed thinkers who were against progressivism, and explicitly anti-Hegelian: these included Schopenhauer, Kierkegaard, Burckhardt, and Nietzsche.[1] It is interesting to note that, of these, Schopenhauer was influential on Santayana and Popper; and Nietzsche on Heidegger, Foucault, and Derrida. So, much of the 20th century opposition to progressivism drew strength from what were similar lines of thought in the 19th century. Also significant is the fact that, now, in the early years of the 21st century, both Schopenhauer and Nietzsche are influential anew, and in many fields as well as that of historical studies.

To conclude: In the West, anti-historicism prevails among leading minds; and because this position includes opposition to progressive historicism, the predominant view is that no moral significance can be attached to history *as a whole.* Certainly history embodies no general confirmation or endorsement of the democratic-humanitarian values held by those in the West who are, arguably, the most morally sensitive. Indeed, the holding of these values is regarded as a basis for *not* endorsing a huge number of actions committed in the past. From this standpoint, much of human history is viewed with profound dissatisfaction, and even horror. The same applies to much which is happening in the present-day world, and which may well continue in the future. There is, then, a foundational opposition between democratic human-itarianism and a great deal of what has constituted, and continues to constitute, human reality.

Since much of that reality does not deliver ethical meaning, the latter has to be supplied, by the morally

---

[1]   For a good summary of the anti-Hegelianism of Schopenhauer and Burckhardt, see Eric Heller, *The Disinherited Mind*: London, Penguin Books Ltd., 1961 (1952), pp. 65–77.

concerned and committed. It is something made, not found, created and not received. These actions take two forms: one is a forthright condemnation of past actions deemed to require it; and the other is stalwart opposition to present behaviour which, again, calls for censure in the strongest terms.

Finally, this call to action has of course nothing to do with the notion of historical laws. While morally affirmative conduct can be regarded as caused, the causes are not laws of history; they are nothing more or less than particular stimuli in particular time-contexts — stimuli which them-selves have particular and local antecedents.

Value-Positions in Dawkins & Darwin

With both the above, as indeed with all commentators on the process of biological evolution, value-positions have to be examined, over and above the purely factual content, or at least the purely factual claims, contained in their work. As previous material in this book has indicated, what is meant by value-position is essentially an attitude: a way of feeling about a fact or alleged fact.

In Richard Dawkins, the attitude expressed toward the general outcome of the evolutionary process is highly, even unreservedly, positive. In the closing pages of his recent book, *The Greatest Show on Earth* (2009), he describes the current forms of life resulting from evolution as, apparently without exception, "most beautiful and most wonderful".[1] Also, these forms exist within a general context of biological activity which includes "stalking, chasing, fleeing, out-pacing, outwitting" (p. 426). Further, the nervous systems of highly complex animals, while in themselves "wonderful", have evolved from the (fully acknowledged) grimness of "the ever-escalating arms races between predators and prey, parasites and hosts", and are all parts of the "greatest" spectacle, that of evolutionary outcome, which the Earth has to offer (p. 426).

It is clear that Dawkins regards what he and many other people see as the harsh and cruel aspects, both of the evolutionary past and the present biological order, as a price worth paying for the existence of the animal beauties and wonders which he claims to be ubiquitous. This position is a

---

[1]   p. 426 of the Black Swan edition of *The Greatest Show on Earth*: London, Black Swan, 2010. All subsequent quotations, unless otherwise stated, will be from this edition of the text, with only page numbers given.

distinctly attitudinal one; and, given his own wording, seems to denote an *aesthetic* stance toward the natural world at least as much as a moral stance. Aesthetic terms abound: in addition to the ones already quoted, there is "intricacy", "complexity", "elegance", and, of course, "show" (as in stage show) (all p. 426). Interest is strongly focused on shape and form, structure and design. To extend an earlier point, the position is that consideration of pain and struggle, which calls chiefly for a *moral* response,[2] should be subordinated, partly at least, to consideration of the aesthetic aspects of the long-term results of that turmoil.

This viewpoint parallels that of Charles Darwin, who is manifestly an inspirational figure for Dawkins. Dawkins quotes Darwin's words in the final paragraph of (the first edition of) *The Origin of Species:*

> Thus, from the war of nature... the most exalted object we are capable of conceiving, namely the production of the higher animals, directly follows. There is grandeur in this view of life... from so simple a beginning, endless forms most beautiful and most wonderful have been, and are being, evolved. (p. 399)

As well as the phrases "most beautiful and most wonderful", which Dawkins has obviously adopted, we have the further aesthetic diction, "exalted object" and "grandeur". As with Dawkins, an aesthetic perspective co-exists with references to harshness and violence.

Further on Dawkins: as said earlier, he does not talk exclusively in aesthetic terms. His concerns are also moral, and he places central moral value on the ability to survive and propagate: a position on which he has been consistent since the publication of his first book, *The Selfish Gene,* in the 1970s. He adopts Darwin's phrase "the higher animals", and uses it precisely as Darwin did to mean those animals best

---

[2]   And, given present-day concerns with animal welfare and even rights, I trust it will not be rejoined that suffering at the sub-human level is of no moral relevance to humans.

able to survive, flourish, and procreate. So, "higher" means 'more able to meet a particular moral criterion', as well as meaning 'more aesthetically pleasing'. In Dawkins and Darwin, there is, then, a combination of moral and aesthetic elements in the value-position espoused.[3]

Now, in comment on the foregoing points: Firstly, issue can be taken with Dawkins' argument about the ubiquity of "the most beautiful" and "the most wonderful". These are of course value-terms, never fact-terms; hence their application can always be disputed. Many people would say they see nothing beautiful in the predatory practices which Dawkins lists on p. 426 (see previous quotation), or in the outcomes of these practices, or in the predators themselves. Indeed, they would claim to find reinforcement for their view in a photograph which, ironically, Dawkins himself provides (illustration 30) of a lioness about to catch up with and devour a kudu.

A prominent spokesperson for their position is actually an English contemporary of Darwin, Tennyson, who famously described nature as "red in tooth and claw". A further spokesperson is another 19th century figure, Schopenhauer, who (see earlier essay) expressed outright horror at the natural order: "everywhere in nature, we see contest, struggle and the fluctuation of victory… This universal conflict is seen most clearly in the animal kingdom…"[4]

A main point emerging from the words of both Tennyson and Schopenhauer is that the pain pervading the animal world, both in the evolutionary past and in the present, needs to be given much more moral attention than either Dawkins or Darwin apportions to it. Their argument

---

[3]   For Darwin, see the following additional vocabulary implying moral approval of "the higher animals": "the strongest" (p. 400), "the vigorous, the healthy and the happy" (p. 401).

[4]   In *The World as Will and Representation, Vol. 1*: trans. E.F.J. Payne, New York, Dover Publications Inc., 1958, pp. 146–7. Incidentally, Schopenhauer, unlike Tennyson, died before he could become acquainted with Darwinism.

that such suffering has been a price worth paying for the evolutionary outcomes it has produced can clearly be challenged. While everybody acknowledges that more complex forms of life (pre-eminently, human beings) have indeed resulted from pain and struggle, not everybody is triumphalist about this fact, or rejoices at it. Some people actually regret the processes by which it has become fact. Thus, their attitude to their own existence as the outcome of these processes is complex, divided, ambivalent, and in some cases leads to ascetic or even renunciatory outlooks. This is obviously a far cry from the Dawkins-Darwin position.

What this viewpoint involves is a questioning, not to be found in either Dawkins or Darwin, of whether it is actually *good* to survive, thrive, and propagate. For those holding this viewpoint who adopt a non-cognitivist position in ethics, there is no objective discovery to be made that survival and procreation are good things. Hence, there is no possibility of taking it for granted that they are. To assert the contrary, as Dawkins and Darwin do, and apparently as a matter of course, requires a justification which they do not supply. Either one must claim to know, objectively, that survival is good, and therefore declare oneself an ethical cognitivist; or one must express an emotional assent to and endorsement of survival, but emphatically go no further than that, thereby showing oneself to be an ethical non-cognitivist. Neither Dawkins nor Darwin is sufficiently explicit on this matter.[5]

Survival is of course a fact, a resultant of evolutionary process, and therefore an 'is' about the world. But, as Hume pointed out, a century before Darwin, no 'ought' can, with philosophical legitimacy, be directly derived from an 'is'. In other words, the fact that something is the case is no indicator that, by virtue of being such, it ought to or should

---

[5]   It is worth noting that all the criticisms of Darwin and Dawkins made in this paragraph apply equally to Spencer, 19th century philosopher of Darwinism, and to Nietzsche, 19th century general commentator on Darwinism.

be the case. An 'ought' has to be argued for, extensively and painstakingly, in a quite separate exercise from the observation of an 'is'.

Index of Names

.